Lecture Notes in Economics and Mathematical Systems

Managing Editors: M. Beckmann and W. Krelle

256

Convexity and Duality in Optimization

Proceedings of the Symposium on
Convexity and Duality in Optimization
Held at the University of Groningen, The Netherlands
June 22, 1984

Edited by Jacob Ponstein

Springer-Verlag
Berlin Heidelberg New York Tokyo

Managing Editors

Prof. Dr. M. Beckmann
Brown University
Providence, RI 02912, USA

Prof. Dr. W. Krelle
Institut für Gesellschafts- und Wirtschaftswissenschaften
der Universität Bonn
Adenauerallee 24–42, D-5300 Bonn, FRG

Editor

Prof. Dr. Jacob Ponstein
University of Groningen
P.O. Box 800, 9700 AV Groningen, The Netherlands

ISBN 3-540-15986-X Springer-Verlag Berlin Heidelberg New York Tokyo
ISBN 0-387-15986-X Springer-Verlag New York Heidelberg Berlin Tokyo

Printed in Germany

Printing and binding: Beltz Offsetdruck, Hemsbach/Bergstr.
2142/3140-543210

This volume contains the papers (except one) presented during the symposium on Convexity and Duality in Optimization, held on June 22nd 1984 at the University of Groningen, The Netherlands, at the occasion of conferring the Honorary Degree in Mathematics and the Sciences to R. Tyrrell Rockafellar, of the University of Washington, Seattle, U.S.A., on June 20th 1984.

TABLE OF CONTENTS

MATHEMATICAL FAITS DIVERS

J.-B. Hiriart-Urruty*)

INTRODUCTION.

This note, which is concerned with faits divers that do not have very much to do with a mathematician's usual activities, is the written version of a talk that served as an introduction to a more mathematical lecture that I had the opportunity to present during the symposium *'Convexity and Duality in Optimization'*, at the occasion of the festivities marking the 370th anniversary of the University of Groningen, The Netherlands.

Before this first visit of mine to Groningen, my knowledge of its university was limited to short discussions regarding the mathematical activities of colleagues and to the fact that the Bernoulli's, in a particular way, were associated with it in the XVIIth century. Looking closer it appears that Groningen and Toulouse show certain similarities, if only because J. Bernoulli had a teaching assignment at Groningen towards the end of the XVIIth century, whereas P. de Fermat was a member of the Parliament of Toulouse a few years before. Besides that, the historical lines find themselves strengthened by the intermediary of Th. J. Stieltjes, whose career, as we will see later on, is closely related to the University of Toulouse. Further, Groningen is the main city in the northern part of The Netherlands. With its around 170000 inhabitants it is the capital of a province carrying the same name. And Toulouse, the ancient capital of Languedoc, is the main city of the Midi-Pyrénées and has around 400000 inhabitants. Both cities have a relatively excentric position with respect to the national capitals and, to a vertain extent, developed independently from the influence of the latter. Moreover, both are marked by the old tradition of their universities, as the University of Toulouse was founded in de XIIIth century and the University of Groningen (first as the Provincial School of Higher Learning for the city and its surroundings) dates back to 1614.

1. THE BERNOULLI'S AND FERMAT. THE BIRTH OF THE FIRST VARIATIONAL PRINCIPLES.

1.1. In three generations, the Bernoulli family produced eight mathematicians, of which the most wellknown are Jacob (1654-1705), his younger brother Johann (1667-1758) and the son of the latter, Daniel (1700-1782). Jacob Bernoulli

*) Translated from the French by J. Ponstein; "..." is a real quotation, '...' is a translated one.

became professor of mathematics at Basle in 1687 until his death; in particular, he is known for his contributions to the theory of probability (notably Bernoulli's law of large numbers). His brother, Johann, started taking courses in medicine before turning to calculus, which he mastered quickly and then applied to various problems in geometry, differential equations and mechanics. In 1695 he was appointed professor of mathematics and physics at Groningen, an assignment that he held until the death of his brother. In 1696 Johann Bernoulli proposed to the scientific community a *'new problem that the mathematicians are invited to solve'*; it is the famous problem of the *brachistochrone: 'given two points A and B in a vertical plane, find the path connecting A and B such that a body subject to its weight would traverse the trajectory between A and B in minimum time'*. Historically, that year 1696 is the beginning of the *calculus of variations*, even though Newton had announced before, in 1687, his *"De motu fluidorum et resistentia projectilium"*).

The problem of the brachistochrone was solved by Leibniz, the two Bernoulli brothers and Newton. In fact, when Newton heard tell of the problem, he solved it immediately and sent his answer back anonymously. When Johann Bernoulli had read it he declared right away that *'one recognizes the lion from the way he attacks'*. Leibniz's solution was based on the technique of approximation curves by polygonal lines. Jakob Bernoulli's solution, finally, based on the principle of Huygens (1629-1695; a Dutch physicist and astronomer, who, as one knows, contributed to Fermat's principle of least time,was more general but also more laborious than that of his brother Johann. The solution proposed by Johann Bernoulli is the most popularized ([1], pp. 25-31, [6] pp. 22-28); among others it uses Snell's law of refraction (W. Snell, 1591-1621, a Dutch astronomer and mathematician, discovered this law in 1621). Mainly as a consequence of having solved the problem of the brachistochrone simultaneously, the relationship between the brothers Jacob and Johann became stormy. It seems that Johann was the one with the more bad temper of the two, not hesitating to show his son the door when he came home with a prize of the French Academy of Sciences, a price he was eager to receive himself. This son, Daniel, was born at Groningen in 1700. After, like his father, having started taking courses in medicine, he turned to mathematics and became a professor at Saint Petersburg before returning to Basle in 1733. This suggests that despite of the difficulties of communication (both in writing and between people) mathematicians of the time moved from one university to another. With his book *Hydrodynamica* Daniel Bernoulli is considered one of the founders of hydrodynamics and the kinetic theory of gases.

1.2. When, in 1981, I left Clermont-Ferrand for Toulouse I followed in the footsteps of the great mathematicians of the XVIIth century, because I left the city of Pascal for that of Fermat.

Beyond doubt Pierre de Fermat (1601-1665) is the greatest mathematician of the XVIIth century, even though his influence was limited by his lack of interest in publishing his findings, that became known via letters to friends and via notes in the margins of books he read. After his studies in law, P. de Fermat became a member of the Parliament of Toulouse in 1631, and he continued this parliamentary career until his death in 1665. In fact, mathematics was his passion and if one is forced to mention three dominant themes in his work, one could cite the following:

- his famous theorems in the theory of numbers, contributions for which he undoubtedly is known best;
- his contributions to the area of the calculus of probability (with Pascal);
- setting up the analytic geometry (with Descartes) and the first variational principles in calculus and optimization.

The *variational principle of Fermat* (1657) and its implications to the law of refraction of light are sufficiently known to consider them any further. The works by Fermat on the *'methods for seeking extrema'* are probably less wellknown. In 1629, (that is thirteen years before the birth of Newton), Fermat conceived his method *"De maximis et minimis"* that could be applied to the determination of values yielding the maximum of the minimum of a function, as well as of tangents of curves, which boiled down to the foundations of differential calculus. Fermat should be considered the initiator of differential calculus, before d'Alembert, Lagrange and Laplace. Fermat's text, communicated by him to one of his colleagues of the Parliament of Bordeaux, came into Descartes' hands not until 1638. Descartes considered it, critized it and, not being friendly towards Fermat, to whom he was already opposed to, declared that the appraised method was false. In fact, Descartes had not understood Fermat's method quite well (it is true that Fermat did not make many efforts to verify his rule!), and the discussion regarding the subject "De maximis et minimis" came to an end by a letter that Descartes sent to Fermat directly, and in which he accepted his rule as being correct.

If one considers the rule proposed by Fermat with a little modern symbolism, one obtains: *'in order to find the value where a function is maximum or minimum: write down $f(a)$, substitute $a+h$ for a, that is to say, write down $f(a+h)$; 'adequalize' $f(a)$ and $f(a+h)$ (adequalize is a term coined by Diophantus to recall that $f(a+h)$ is only "almost equal" to $f(a)$); in this adequality only consider the terms of the first order in h; divide by h; then setting the result equal to zero, one obtains an equation in a from which a*

has to be solved'. In other words, one way of using the derivative f' for determining the extrema is by seeking the solutions of $f'(a) = 0$. Certainly, Fermat used this rule (and hence a corresponding differential calculus) only for classes of functions that where useful to him, that is to say, the algebraic functions of a real variable, the trigonometric functions and compositions thereof. Notwithstanding the attacks aimed at him, Fermat kept full confidence in his method that, as he put it, *'is never mistaken and can be extended to very nice questions'*. Furthermore, it seems that he knew quite soon how to distinguish the various possible types of extrema by neglecting the terms of the 3rd order in his expansion of f(a+h); *'the main point is'*, as he put it, that the terms of the 2nd order are *'of greater importance'* than the terms of the highest power beyond the 2nd order. In Fermat's explanations a notion of 'differential' appears, hidden in the idea of an increment of a quantity that is a function of the variation of another quantity that one let 'disappear'. The delicious French expression "quantité évanescente", dear to analysts of the XVIIth and XVIIth century, is now out of use for already a long period of time. In English the related expression "vanishing term" has, however, been preserved.

Fermat also proposed a method for finding tangents of a curve of which the Cartesian equation is given. We will not discuss that here.

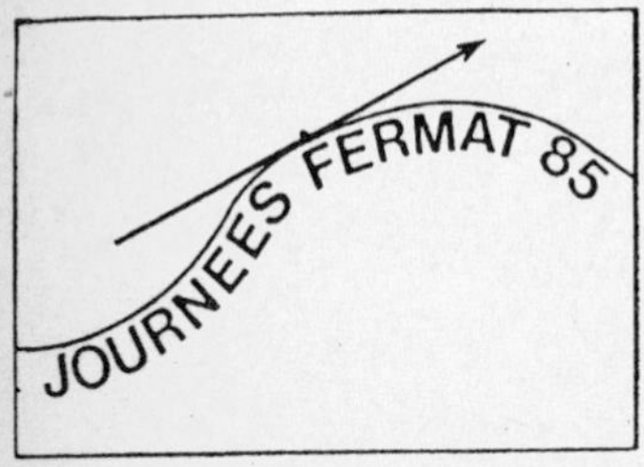

The variational principle of Fermat, called, in the language of optimization, *necessary optimality condition of the 1st order*, and his method for determining tangents of a curve, led me to propose the diagram to the left for the Journées FERMAT 1985 dedicated to the 'Mathematics of Optimization'.

In the badly organized scientific world of the XVIIth century (recall that the Academy of Sciences of Paris was founded not until 1666!), the circulation of scientific writings had an almost confidential character. Some, like Mersenne and Carcavi (who admired Fermat) had the zeal and a strong disinterest to act as intermediates between the scientists. Descartes, Desargues, Roberval in Paris, Torricelli in Italy, Van Schooten, who introduced the young Huygens at Leiden, thus received copies of the treatises of Fermat from Marsenne.

Among the optimization problems to which Fermat's name is attached (there are several of them!), there is one that deserves to be considered in more detail, because it is concerned with a problem of *convex nondifferentiable optimization*. The problem reads as follows: *given 3 noncollinear points in the plane, A, B, C, find the point P that minimizes the sum of the distances of P to A, B and C.*

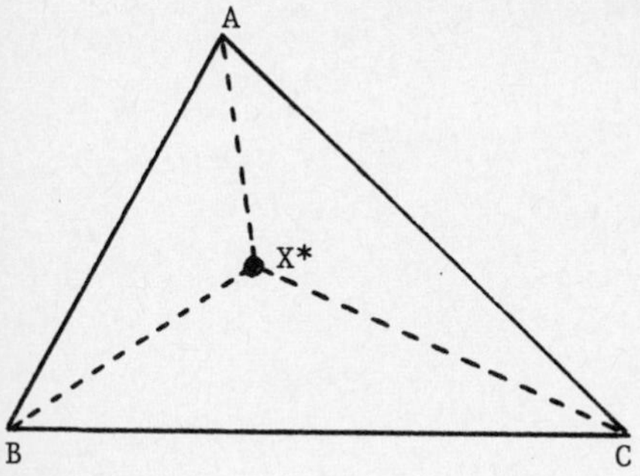

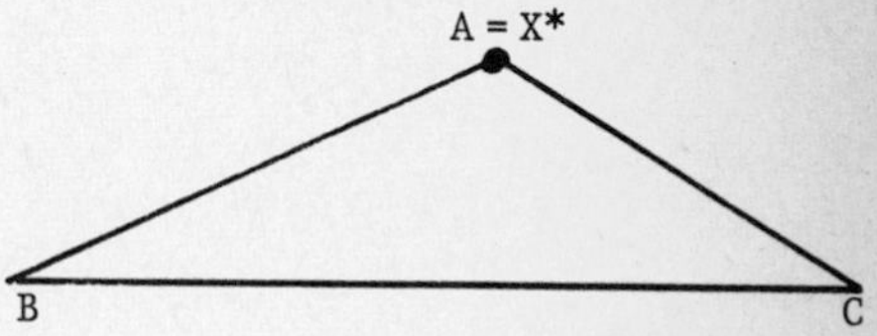

This problem can certainly be solved by simple geometric considerations. From the point of view of optimization, observe that the function to be minimized, i.e.

$$f: X \to f(X) = |X - A| + |X - B| + |X - C|,$$

(where $X \in \mathbb{R}^2$ is the vector whose components are the coordinates of a running point P) is *coercive* ($\lim_{|X|\to+\infty} f(X) = +\infty$) and *convex*. Only at the points A, B and C f is nondifferentiable. If one knows that the solution X^* is to be found in the interior of the triangle ABC, which then can be seen by a rather simple reasoning, one can write:

$$Df(X^*) = 0 \text{ (Fermat's rule).}$$

If this is not the case, the necessary and sufficient optimality condition can be written as:

$$(*) \quad 0 \in \partial f(X^*),$$

where ∂f indicates the *subdifferential* of f. This example shows quite well to students the distinction between using either Df or ∂f, depending on whether or not one has supplementary information regarding the candidates for the optimal solution X^* (information that one can deduce from the relative position of A, B and C). Solving (*), or

$$0 \in \partial(|.-A|)(X^*) + \partial(|.-B|)(X^*) + \partial(|.-C|)(X^*),$$

one obtains this: *if the 3 angles of the triangle ABC are less than 120°, then the point X^* where the minimum of f is assumed is the one from which one can* *see all 3 sides under an angle of 120° (this point is sometimes called the Torricelli point or the Steiner point); and if one of its angles is at least 120°, then X^* is the corresponding vertex of the triangle.*

It is wellknown that this problem has many generalizations in operations research. Given n distinct points A_i in the plane, find X minimizing

$$f(X) = \Sigma_{i=1}^{n} \sigma_i(|X-A_i|),$$

where $\sigma_i : \mathbb{R}_+ \to \mathbb{R}_+$ is a convex, increasing function, measuring the cost associated with the distance from A_i (think of the problem of finding the localization of an heliport serving n positions). Minimizing f reveals the usual techniques of convex optimization.

In a simple version, this leads me to pose the next exercise to mathematics students during their first year: given n distinct points in $\mathbb{R}$, find the point(s) minimizing:

$$f(x) = \Sigma_{i=1}^{n} |x-a_i|, \quad x \in \mathbb{R}.$$

The *existence* of a solution, the *convexity* of f, and the *characterization* of the solutions (whether or not using the techniques of elementary differential calculus) are just as many questions that may lead students astray. In any case it is a good exercise, which one inevitable will stumble upon when making ones first steps into statistics.

1.3. According to the historians of mathematics, one of the types of problem that motivated the creation of calculus in the XVIIth century, was to find the maximum or minimum value of a given criterium. Through the calculus of variations this has led to a modern branch which became more and more separated from it: *the theory of optimal control*. The interested reader will read with pleasure the book by Alexeev, Tikhomirov and Fomine, *Commande optimal*, containing an abundance of examples.

Following the variational principle of Fermat, several other variational principles were discovered, finding their origin in mechanics. In this respect, I like to parody a wellknown song by saying: *'There is always an extremum principle of some sort'*. Let us conclude by citing Euler ([1, p. 25]): *'Nothing goes on in the world that has not the meaning of a certain maximization or a certain minimization'*. Maximization of profit? Of the happiness of people? I leave this to the imagination of the reader.

Before jumping to the XIXth century, let us make a quick stop at the XVIIIth century to mention A.-M. Legendre, born at Toulouse in 1752, who mathematicians inspired by the calculus of variations will not fail to know.

2. THOMAS-JAN STIELTJES (1856-1894).

If one had to choose a mathematician whose name is remembered in the mathematical communities of both France and The Netherlands, one would not hesitate to select Th. J. Stieltjes. The mathematical career of Stieltjes is closely related to that of French mathematicians toward the end of the century, in particular when he was a professor at the University of Toulouse during almost ten years, and got acquainted to French mathematicians such as C. Hermite. One of the detours of his career brought him to the University of Groningen, be it under unfortunate circumstances.

Th. J. Stieltjes was born at Zwolle (The Netherlands) in 1856. In Delft he finished secondary school and entered the Technological University of this city. Although his studies did not cause difficulties, he failed to pass the final exams two years in succession and did not get his diploma in engineering. This failure would well have its consequences in the future. When at first being employed as an 'astronomical calculator' at the observatory at Leiden, he uses his time to saturate his passion for mathematics. Finding his work as an astronomer to hard, in 1878 he ponders to quit and go to the other side of the Atlantic to study under Sylvester. But the director of the observatory who understands his inclination towards theoretical work, relieved him from certain tasks. His productivity intensified.

It is November 1882 when Stieltjes, in Paris, makes contacts with Hermite. This was the beginning of a long correspondence that lasted until Stieltjes' death. Hermite was one of the greatest mathematicians of his time, but certainly he was the greatest writer of letters. He maintained relations with mathematicians all over Europe, spending his time and showing his ideas to whose who asked for advice or submitted to him the first set-up of their works. Over 430 letters were exchanged between Stieltjes and Hermite. Some of them are savoury and show the evolution of ideas, as well as the reservations of both mathematicians.

In November 1883, Stieltjes plans to abandon astronomy completely. His importance is now recognized by mathematicians close by in The Netherlands, so he does not hesitate to pose his candidature for a chair at the University of Groningen. He does not get the job and in his letter to Hermite piquant passages about this are not missing (March 14th, 1884): *'For the vacant position the Faculty at Groningen has given the highest priority to me, but the Minister has nominated one of the others. Probably, the reason will have been that because there was no opportunity at all to follow the ordinary way, I have no university degree at all ...'.*

Embittered by this misfortune, Stieltjes nevertheless takes up his work again and in 1885 decides to leave his country definitely and to live in Paris.

Then the course of events accelerates. Hermite proposed to Stieltjes to present a doctoral thesis to the Faculty of Sciences of Paris, that, as he wrote him, *'would open the doors of our superior educational institutions for you, and certainly would lead you, each time if such a situation would be acceptable to you, to become a professor at some Faculty of Sciences ...'*. The thesis, entitled 'Study of some semi-convergent series' was defended on June 30, 1886. Hermite was the chairman of the examining board, Darboux and Tisserand were the examiners.

On October 26, 1886 Stieltjes learns that he is charged with a course in mathematics at the Faculty of Sciences at Toulouse. Immediately he moves to Toulouse, with his family, to stay there until his death in 1894. The ensueing scientific production of Stieltjes is as one knows it (cf. [3,7]). The dialogue with Hermite continues, it could happen that Stieltjes wrote two letter a day to Hermite! Among the answers of Hermite there is one, well-known, that is full of taste for those, like me, who are interested in the analysis and optimization of nondifferentiable functions: *'With fright and horror, I turn my back to this lamentable plague of continuous functions that have no derivative ...'*.

Several times Stieltjes received an award of the Academy of Sciences at Paris for his works. In 1892, he became a candidate for the succession of O. Bonnet's position at the Academy. He was classified as number two, ex aequo with H. Poincaré; P. Appel being classified as number one.

By the end of May, 1894, Stieltjes can announce to Hermite that he has finished the manuscript of his memoir entitled 'Investigation into the continued fractions'. Notably, Stieltjes here introduces the concept of the 'distribution of positive mass' over the reals, which led him to what since then are called the Stieltjes measures and the Stieltjes integral.

Physically exhausted Stieltjes passed away on December 31, 1894, a little over 38 years old.

In 1966, when two class rooms were reserved for mathematicians in the new localization of the Faculty of Sciences at Toulouse in Rangueil, naturally the names of Fermat and Stieltjes were given to them. That does not alter the fact that after a one year course in Stieltjes theatre students have only an evasive idea of 'Thomas-Jan Stieltjes' unusual fate' ([5]).

REFERENCES

[1] ALEXEEV, V., TIKHOMIROV, V. et FOMINE, S., Commande optimale, Editions Mir, 1982.

[2] BAYEN, M., Histoire des universités, Collection "Que saus-je?" n° 391, Presses Universitaires de France, 1973.

[3] COSSERAT, E., Notice sur les travaux scientifiques de T.-J. Stieltjes, Annales de la Faculté de Sciences de Toulouse 9, 1895, 1-64.

[4] HURON, R., L'aventure mathématique de Fermat, XXIe Congrès de la Fédération des Sociétés Académiques et Savantes Languedoc - Pyrénées - Gascogne, 15-16 Mai 1965.

[5] HURON, R., Le destin hors série de Thomas-Jan Stieltjes, Mémoires de l'Académie des Sciences Inscriptions et Belles-Lettres de Toulouse, Vol. 136, 1974, 93-125.

[6] SIMON, G.F., Differential equations with applications and historical notes, Mc Graw-Hill, Inc., 1972.

[7] Oeuvres complètes de Th. J. Stieltjes, Tomes 1 et 2, plubliés sous les auspices de l'Académie d'Amsterdam, P. Noordhoff, Groningen, 1914 et 1918.

[8] TANNERY, P., et HENRY, C., Oeuvres de Fermat, Tomes 1 à 4 et Supplément, Gauthier-Villars, Paris, 1891, 1894, 1896, 1912 et 1922.

[9] TERJANIAN, G., Fermat et son arithmétique, Séminaire d'Histoire des Mathématiques, Février 1980.

MONOTROPIC PROGRAMMING: A GENERALIZATION OF LINEAR PROGRAMMING AND NETWORK PROGRAMMING

R. Tyrrell Rockafellar*

Duality schemes have been developed for almost all types of optimization problems, but the classical scheme in linear programming has remained the most popular and indeed the only one that is widely familiar. No doubt this is due its simplicity and ease of application as much as its close connection with computation, which is a property shared with other dualities.

Linear programming duality has an appealing formulation in terms of a matrix or tableau in which each row or column corresponds to both a primal variable and a dual variable. A role assigned to one of these paired variables automatically entails a certain role for the other. In the usual way of thinking, decision variables in one problem correspond to slack variables in the other. The only flexibility is whether a decision variable is nonnegative or unrestricted, in accordance with which the corresponding slack variable is tied to an inequality or an equation.

A vast extension of this approach to duality is possible without sacrificing the sharpness of results or even the "concreteness" of representation. The extension is achieved by admitting far more general roles for the variables in a primal-dual pair. To appreciate how this is possible, it is necessary first to free oneself from the idea that there is something inherent in a variable being either a decision variable or a sort of slack variable. Such a distinction is actually an impediment even in standard linear programming, because it relates more to the initial tableau being used than to something inherent in the variables themselves. Whether a variable is "independent" or "dependent" in the expression of a problem and its constraints can change when pivoting transformations are applied to the tableau.

In the generalized duality scheme which we call *monotropic programming* the "role" of a variable is something apart from its incidental position in a tableau. It is given by specifying both an *interval* (the range of values permitted for the variable) and a *cost expression* defined over the interval. The cost expression is a convex function of the variable (possibly linear or piecewise linear) and it might be everywhere zero on the interval, in which case one *could* appropriately think of the variable as a slack variable in a broadened sense.

The pair consisting of an interval and a convex cost expression on that

---

*Supported in part by a grant from the National Science Foundation at the University of Washington, Seattle.

interval can be identified with a so-called proper convex function on the real line. Subject to a minor regularity condition ("closedness", a property referring to endpoints), such a function can be dualized: the *conjugate* function furnishes the interval and associated cost expression defining the "role" of the dual variable.

Conjugates of convex functions of a single variable can readily be constructed by a process of generalized differentiation, inversion and reintegration. This feature gives monotropic programming duality a potential for much wider use in applications than other, more abstract forms of duality. Another interesting feature is the way that monotropic programming associates with each primal-dual pair of variables a certain "maximal monotone relation". The prototypes for such relations in ordinary linear programming are the complementary slackness relations used in characterizing optimality. In network programming, however, the relations have an importance all of their won. They describe the combinations of flow and tension (potential difference) that can occur in the various arcs of the network more or less like generalizations of Ohm's Law, which in classical electrical theory corresponds to an arc representing an ideal resistor.

The theory of monotropic programming duality is presented in complete detail in the author's 1984 book [1]. Nevertheless a briefer introduction to the subject may be helpful, because the results and even the basic ideas are not yet widely known but very applicable. This is the justification for the present article.

In order not to encumber the exposition, proofs are omitted. Nothing is said about the history of the subject, the theories that are related to it, or to people who have made major contributions. All that can be found in [1].

1. LINEAR SYSTEMS OF VARIABLES.

A fundamental concept in monotropic programming is that of a "finite collection of real variables which are interrelated in a homogeneous linear manner". Let us denote the variables by x_j, $j \in J$, where J is some finite index set (such as $\{1,\dots,n\}$ or $\{0,1,\dots,n\}$ or $\{1,\dots,n\} \times \{1,\dots,m\}$; flexibility in this respect is helpful). We can think of vectors

$$x = (\dots,x_j,\dots) \in R^J$$

as corresponding to an assignment of a real number value to each variable. Such vectors can be added or multiplied by scalars as usual:

$$(\dots,x_j,\dots) + (\dots,x'_j,\dots) = (\dots,x_j+x'_j,\dots)$$
$$\lambda(\dots,x_j,\dots) = (\dots,\lambda x_j,\dots)$$

In this sense R^J is a vector space (identifiable with R^n when $J = \{1,\dots,n\}$). To say that our variables x_j are related in a homogeneous linear manner is to say that the value vectors x we are interested in form a subset C of R^j with the property

$$x \in C,\ x' \in C \ \Rightarrow\ x + x' \in C,$$

$$x \in C,\ \lambda \in R \ \Rightarrow\ \lambda x \in C.$$

In other words, C is (linear) subspace of R^J.

Formally speaking, then, a *linear system of variables* is simply the designation of a subspace C of R^J for some finite index set J. Often this subspace is described for us initially by a set of homogeneous linear equations

$$(1) \qquad \sum_{j \in J} e_{ij} x_j = 0 \text{ for } i \in I,$$

where I is some other index set, but this description is not unique and can be re-expressed in many different ways. (The restriction to homogeneous linear relations at this stage is merely a theoretical device, but an important one. General linear constraints will be handled below by specifying for each j a real interval C_j in which x_j must lie. Included in this is the possibility that some of these intervals consist of a single point, so that the corresponding x_j's must take on fixed values. Then (1) turns into a system of inhomogeneous linear equations in the remaining x_j's.)

A common way for a linear system of variables to be described is through a set of equations of the special form

$$(2) \qquad \sum_{l \in L} a_{kl} x_l = x_k \text{ for } k \in K,$$

where L and K are separate index sets, J = K ∪ L. These could, of course, be written as

$$-x_k + \sum_{l \in L} a_{kl} x_l = 0 \text{ for } k \in K$$

in order to fit the pattern of (1). Very useful in this connection is the *tableau notation* for (2) in Figure 1. When a linear system of variables is given in this way initially, the variables x_l for l ∈ L may be thought of as "inputs" and the variables x_k for k ∈ K as "outputs". From a mathematical point

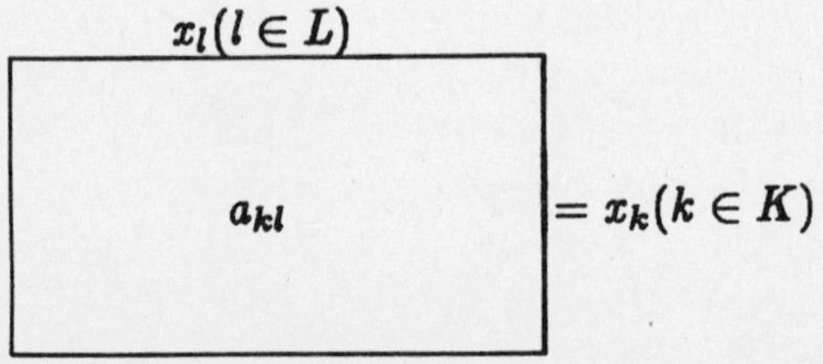

Figure 1.

of view, however, such a distinction is of little importance.

In fact *any* linear system of variables can be represented as in (2) in a *multiplicity of ways*. We speak of these as *Tucker representations* of the system or of the subspace $C \subset R^J$. Specifically, one can pass from (1) to (2) by solving for a maximal set of x_j's (corresponding to some index set $K \subset J$) in terms of the remaining x_j's (corresponding to the complementary index set $L = J \backslash K$).
Thus there is a one-to-one correspondence between Tucker representations of the given C and certain partitions of J into a pair of subsets, K and L. Obviously there are many (but only *finitely* many) such representations, and they all yield tableaus of the same size: the number of elements of L must equal the dimension of C in R^J.

It is possible to pass from any one Tucker representation to any other by a series of *pivoting transformations* of the tableau, each such transformation involving an exchange of an index k_0 in K with an index l_0 in L. This is a central idea in computational procedures in monotropic programming.

2. MONOTROPIC PROGRAMMING PROBLEMS.

We suppose now that we are given a linear system of variables, designated by a certain subspace $C \subset R^J$, and also for each index $j \in J$ a *closed proper convex* function f_j on R. We denote by C_j the (nonempty) interval of R where f_j has finite values (the set dom f_j in convex analysis); more about this in a moment. The corresponding *monotropic programming* problem that we shall be concerned with is

(P) $$\text{minimize } \sum_{j\in J} f_j(x_j) =: F(x) \text{ over all}$$
$$x = (\ldots,x_j,\ldots) \in C \text{ satisfying } x_j \in C_j \text{ for all } j \in J.$$

Note that in terms of any Tucker representation of C this takes the form

(P') $$\text{minimize } \sum_{k\in K} f_k(x_k) + \sum_{l\in L} f_l(k_l)$$
$$\text{subject to } x_l \in C_l \text{ for all } l \in L,$$
$$x_k = \sum_{l\in L} a_{kl}x_l \in C_k \text{ for all } k \in K.$$

Here we recall that a *proper convex* function on R is a function defined on *all* of R with values that are real numbers (i.e. finite) or $+\infty$ (but not *everywhere* $+\infty$) and such that the usual convexity inequality is satisfied. For such a function f_j, the set of points where the value of f_j is not $+\infty$ necessarily forms a nonempty interval, which we are denoting here by C_j. *Closedness* of f_j is a mild semicontinuity condition on the behavior of f_j at the endpoints of C_j (to the extent that these are finite): the value of f_j at such an endpoint must coincide with the limiting value obtained as the endpoint is approached from within C_j. (This allows for the possibility that $f_j(x_j) \to +\infty$ as x_j approaches a finite endpoint of C_j from within C_j; thus the closedness of f_j does not require the closedness of C_j.)

For readers unaccustomed to dealing with $+\infty$, it is essential to realize that the introduction of $+\infty$ is merely a notational device for the representation of constraints which happens to be very useful in theory, particularly in understanding duality. Every pair C_j, f_j, consisting of a nonempty real interval C_j and an arbitrary finite-valued convex function f_j on C_j in the traditional sense, can be identified uniquely and unambiguously with a certain proper convex function on R: simply regard $f_j(x_j)$ as $+\infty$ for every $x_j \notin C_j$.

An x that satisfies $x \in C$ and the interval constraints $x_j \in C_j$ in (P) (or the corresponding conditions in (P')) is said to be a *feasible solution* to our problem, of course. The feasible solutions form a convex subset on R^J on which the objective F is a finite convex function. (Observe that $F(x) < \infty$ if and only if $f_j(x_j) < \infty$ for all j, or in other words, $x_j \in C_j$ for all j. Thus the interval constraints in (P) would be implicit in the minimization even if we had not listed them, which we did for emphasis.)

As represented in the form (P'), our problem could be viewed as one in terms of the variables x_l alone: a certain convex function on R^L which is *preseparable* (expressible as a sum of linear functions of the variables x_l composed with convex functions) is to be minimized subject to a system of linear constraints. To adopt this view strongly, however, would be to miss one of the main features of monotropic programming. Here we are referring to the fact that the representation (P') is in no way unique, and by passing between various such representations in terms of pivoting we hope to gain computational advantage and insight into the underlying problem (P).

3. CATEGORIES OF MONOTROPIC PROGRAMMING.

Many cases of problem (P) are of interest and serve to illuminate the scope of monotropic programming. To look at a "degenerate" example first, suppose that every function f_j is just the *indicator* of a closed interval C_j:

$$(3) \qquad f_j(x_j) = \delta(x_j|C_j) = \begin{cases} 0 & \text{when } x_j \in C_j, \\ \infty & \text{when } x_j \notin C_j. \end{cases}$$

Then we have a *pure feasibility problem*; the objective F(x) has value 0 for all feasible solutions x, and the problem reduces simply to finding such a feasible x, any one at all.

More generally, certain of the function f_j, but not all, may be indicators as in (3). These functions then serve only to represent certain constraints $x_j \in C_j$. They make no contribution to the 'cost' $F(x)$ of a feasible solution x.

These ideas can be made clearer by considering how *linear programming problems* of the standard sort fit the model of monotropic programming. In the case of such problems we imagine the linear system of variables to be given initially in terms of relations of type (2), so that (P') is the form to aim at. Suppose the linear programming problem is

$$\text{minimize} \sum_{l \in L} c_l x_l \text{ subject to } x_l \geqq 0 \text{ for all } l \in L \text{ and}$$

$$(4) \qquad \sum_{l \in L} a_{kl} x_l \quad \begin{matrix} \geqq b_k & \text{for } k \in K_+ \\ = b_k & \text{for } k \in K_0 \\ \leqq b_k & \text{for } k \in K_- \end{matrix}$$

where L, K_+, K_0 and K_- are separate index sets. We can identify this problem with (P') in the case of $K = K_+ \cup K_0 \cup K_-$ and

$$(5) \qquad f_l(x_l) = c_l x_l + \delta(x_l | C_l), \quad f_k(x_k) = \delta(x_k | C_k),$$

where $C_l = [0,\infty)$ for all $l \in L$ and

$$(6) \qquad C_k = \begin{matrix} [b_k,\infty) & \text{for } k \in K_+, \\ [b_k,b_k] & \text{for } k \in K_0, \\ (-\infty,b_k] & \text{for } k \in K_-. \end{matrix}$$

Here it would be easy to introduce upper bounds on the variables x_l: take C_l to be an interval of the form $[0,d_l]$. Another extension would be to allow constraints like $\sum_{l \in L} a_{kl} x_l \leqq b_k$ to be violated, but at a penalty. If the penalty is linear, this would correspond to taking f_k as in Figure 2 and analogously for indices k in K_0 or K_+. Then the objective F would no longer be linear, but piecewise linear.

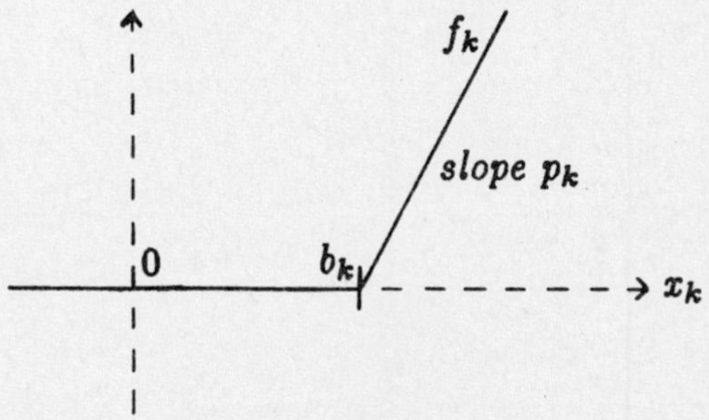

Figure 2.

In general we can identify *piecewise linear programming* as the branch of monotropic programming where each of the functions f_j is piecewise linear relative to C_j with finitely many pieces (i.e. f_j is a polyhedral convex function on R in the terminology of convex analysis). Similarly, if each f_j is piecewise quadratic we have *piecewise quadratic programming*, which can be shown in particular to encompass all of convex quadratic programming, and which also allows for constraint penalties as in Figure 2 but with quadratic portions.

A noteworthy feature of monotropic programming that will be viewed below is the duality which is possible in these categories of problems. For instance, the dual of a piecewise linear problem will be another piecewise linear problem which can readily be constructed. The duality theory of linear programming itself is, in contrast, relatively cumbersome and limited. The dual of a *linear programming problem in the general sense*, i.e. the case of (P) or (P') where each f_j is linear (we should really say "affine") relative to C_j, can be obtained in the traditional framework only by reducing first to a standard type of linear programming problem through modification of the constraints by the introduction of auxiliary variables. This is a drawback in particular for linear programming problems with upper bounds, and the case of linear constraint penalties is even worse. In the context of monotropic programming, the dual of a linear programming problem in the general sense will be piecewise linear rather than linear, but it can be generated directly.

4. NETWORK PROGRAMMING AS A SPECIAL CASE.

A network, or directed graph, as shown in Figure 3, is defined mathematically in terms of finite sets I and J, comprised of the *nodes* i and *arcs* j of the network, and the *incidence matrix*

$$(7) \qquad e_{ij} = \begin{cases} 1 \text{ if } i \text{ is the initial node of the arc } j, \\ -1 \text{ if } i \text{ is the terminal node of the arc } j, \\ 0 \text{ if } i \text{ is not a node of } j. \end{cases}$$

(Arcs that start and end at the same node are excluded.)

For each arc $j \in J$, let x_j denote the amount of *flow* in j (a positive quantity being interpreted as material moving in the direction of the arrow that represents j). The linear system we are interested in consists of the variables x_j, $j \in J$, as related by the conditions

$$\sum_{j \in J} e_{ij} x_j = 0 \text{ for all } i \in I.$$

These conditions are Kirchoff's node laws. They say that at each node i, what enters equals what leaves, or in other words, flow is conserved. The vectors

$x = (\ldots, x_j, \ldots) \in R^J$ satisfying these conditions are called *circulations*, and the subspace C that they form is the *circulation space* for the network.

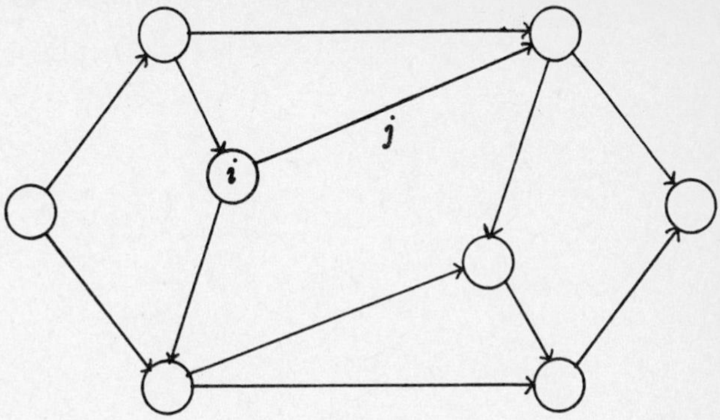

Figure 3.

The possible Tucker representations (2) in this case correspond one-to-one with the arc sets K that are *spanning trees* for the network. Pivoting from one such representation to another can be carried out "combinatorially" through the manipulation of such trees and their associated cuts and circuits, instead of numerical operations on the coefficients in the tableau.

What then is the interpretation of the monotropic programming problem (P)? For each arc j the flow x_j is restricted to a certain interval C_j and assessed at a certain cost $f_j(x_j)$ (possibly zero). Subject to these interval constraints, one seeks a circulation x that minimizes total cost.

The choice of the intervals C_j in such a problem can reflect restrictions on the direction of flow as well as its magnitude in the arc j. Thus for $C_j = [0,\infty)$ we simply have the condition that the flow in j must be from the initial node to the terminal node, whereas for $C_j = [0,c_j]$ the flow must in addition be bounded in magnitude by c_j. Similarly for $C_j = (-\infty,\infty)$ and $C_j = [-c_j,c_j]$: in the first instance there is no restriction on x_j whatsoever, whereas in the second the direction is unrestricted but $|x_j| \leqq c_j$. The case of a single point interval $C_j = [c_j,c_j]$ corresponds to a preassigned value for the flow in the arc j namely, $x_j = c_j$.

The relationship between this version of network programming and other, more traditional modes is clarified in Figure 4. Problems for the network in Figure 3 that might ordinarily be conceived in terms of flows are *not* necessarily conserved at every node are represented in terms of circulations in the augmented network which has a distribution node $\bar{i}$ (a "ground" node in electrical theory). The flow $x_{\bar{j}}$ in the arc $\bar{j}$ in Figure 4 that connects this node $\bar{i}$ to one of the other nodes i corresponds in Figure 3 to an amount of material entering the network at i (positive or negative). Thus a requirement $x_{\bar{j}} = [c_{\bar{j}}, c_{\bar{j}}]$ in this case could be interpreted as specifying the amount entering the network in Figure 3 at i. (If $c_{\bar{j}} > 0$, i would be a supply point, whereas if $c_{\bar{j}} < 0$, i would be a demand point; if $c_{\bar{j}} = 0$, i would be neither.) More general conditions $x_{\bar{j}} \in C_{\bar{j}}$ in Figure 4 could be interpreted as allowing for a certain range of supply or demand at i in Figure 3.

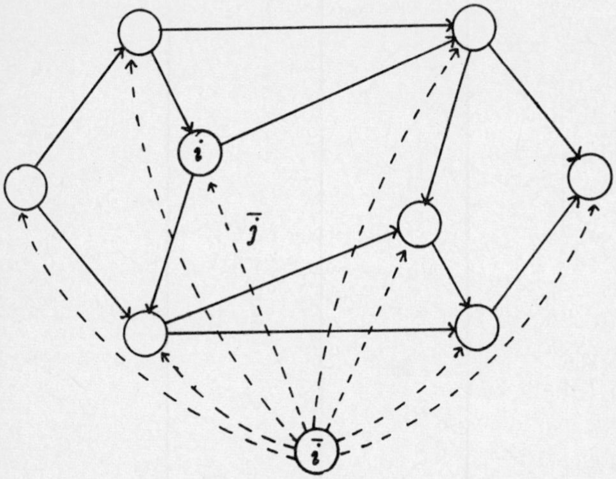

Figure 4.

From these considerations it is evident that monotropic programming problems for flows in networks are *generalized transportation problems* with possibly non-linear costs. In the pure feasibility case they are generalized distribution problems connected with the satisfaction of various requirements of capacity, supply and demand.

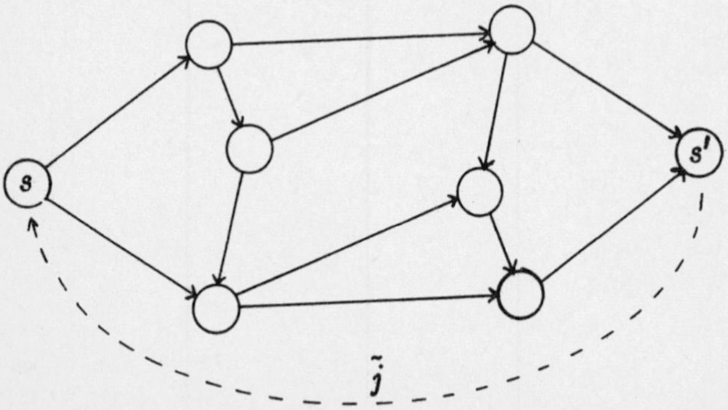

Figure 5.

Even the max flow problem and its generalizations fit this mold. Figure 5 indicates the modified network that would correspond to maximizing in the network of Figure 3 the flow from a certain node s to another node s' subject to capacity conditions $x_j \in C_j$. Over all feasible *circulations* in the modified network, we seek to maximize the flow in the "feedback" arc $\tilde{j}$ (or minimize its negative.) This is the monotropic programming problem that corresponds to taking

$$(8) \qquad \begin{aligned} & f_{\tilde{j}}(x_{\tilde{j}}) = -x_{\tilde{j}} \quad (C_{\tilde{j}} = (-\infty,\infty)), \\ & f_j(x_j) = \delta(x_j | C_j) \text{ for all other arcs } j. \end{aligned}$$

Other important classes of monotropic programming problems for flows in networks involve linear systems of variables more general than the circulation system so far described. For example, there are problems for *network with gains*, where the amount flowing in the arc j can be amplified or attenuated by a certain factor. (The incidences in (7) are replaced by more general numbers.) Such problems too lend themselves to combinatorial rules of pivoting in the manipulation of Tucker tableaus. Also included are *traffic problems* (or *multicommodity flow problems*), where several types of flow may be present in each arc and these compete for the available capacity.

5. DUAL LINEAR SYSTEMS AND PIVOTING.

Suppose we have a linear system of variables x_j, $j \in J$, corresponding to a subspace $C \subset R^J$. The *dual* linear system is comprised of the variables v_j, $j \in J$, under the restrictions

$$(9) \qquad \sum_{j\in J} v_j x_j = 0 \text{ for all } x = (\dots,x_j,\dots) \in C.$$

This system therefore corresponds to the subspace

$$(10) \qquad D = \{v = (\dots,v_j,\dots) \mid v \bullet x = 0 \text{ for all } x \in C\} = C^{\perp}.$$

The dual of the dual system is the primal system: $D^{\perp} = C^{\perp\perp} = C^{\perp}$.

What seems at first quite remarkable about the dual linear system, despite an elementary proof, is the fact that its Tucker representations correspond one-to-one with those of the primal system in a simple way. Suppose the primal linear system is given a Tucker representation as in (2), where J is partitioned into index sets K and L. Substituting this into (9) we get the condition that

$$0 = \sum_{k\in K} v_k [\sum_{l\in L} a_{kl} x_l] + \sum_{l\in l} v_l x_l$$

$$= \sum_{l \in L} [v_l + \sum_{k \in K} v_k a_{kl}] x_l$$

for all possible values of x_l. In other words, the dual linear system is characterized by the relations

$$(11) \qquad v_l = - \sum_{k \in K} v_k a_{kl} \text{ for } l \in L.$$

This is a Tucker representation in which the roles of K and L are reversed and the coefficient matrix is the negative transpose of the earlier one.

The paired Tucker representations of the primal and dual linear systems can be combined in a single tableau as in Figure 6. Here the indices $k \in K$ correspond to

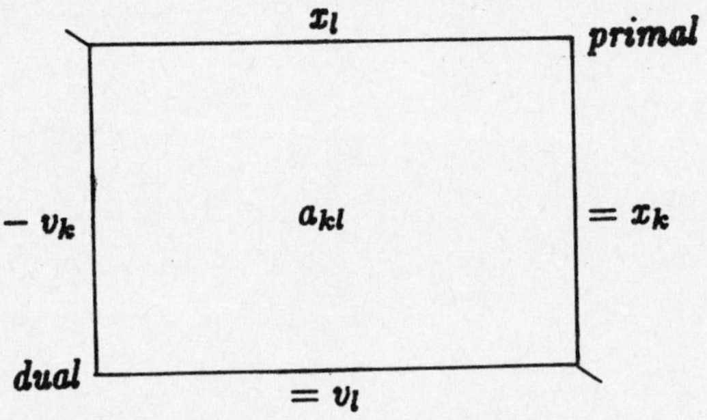

Figure 6.

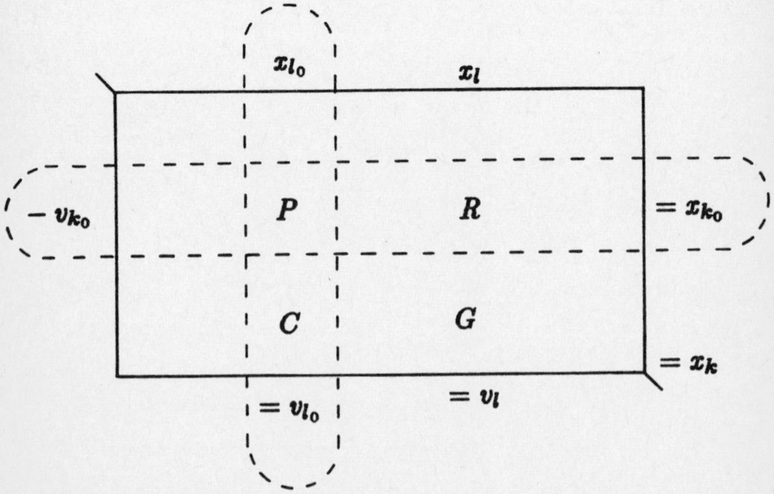

Figure 7.

rows giving equations for the primal system, whereas the indices $l \in L$ correspond to columns giving equations in the dual system. From any Tucker representation of the primal system, written in this manner, one can immediately read off the corresponding Tucker representation of the dual system, and vice versa.

It follows from these observations that pivoting in the two systems can be done in tandem. In a typical pivoting transformation, an index $k_0 \in K$ is exchanged with an index $l_0 \in L$. The (k_0, l_0) entry in the tableau is called the *pivot*, and its row and column are the *pivot row* and *pivot column*; see Figure 7, where P denotes the pivot, R and C denote elements in the pivot row and column other than the pivot, and G is a general entry not in the pivot row or column (but in the same row as C and the same column as R). After the transformation the tableau is as in Figure 8.

The numerical formulas for the transformed coefficients are

$$(12) \qquad P' = 1/P, \; C' = C/P, \; R' = -R/P, \; G' = G - CR/P.$$

One must always remember, however, that in important cases such as occur in network programming such formulas for the coefficients can be by-passed, because it is possible to store the Tucker representations combinatorially in terms of spanning trees and generate particular coefficients a_{kl} from this as needed. In other cases, for instance in traffic problems, a decomposition is possible in which pivoting is carried out partly by numerical formula and partly by combinatorial techniques.

Each variable x_j in the primal linear system is paired with a variable v_j in the dual linear system, and in applications this pairing usually has a natural significance. The case of network programming furnishes a memorable illustration. In that case the space C consists of the flow vectors $x \in R^J$ that satisfy the homogeneous equations (1) for the incidence matrix (7). In geometric terms we therefore

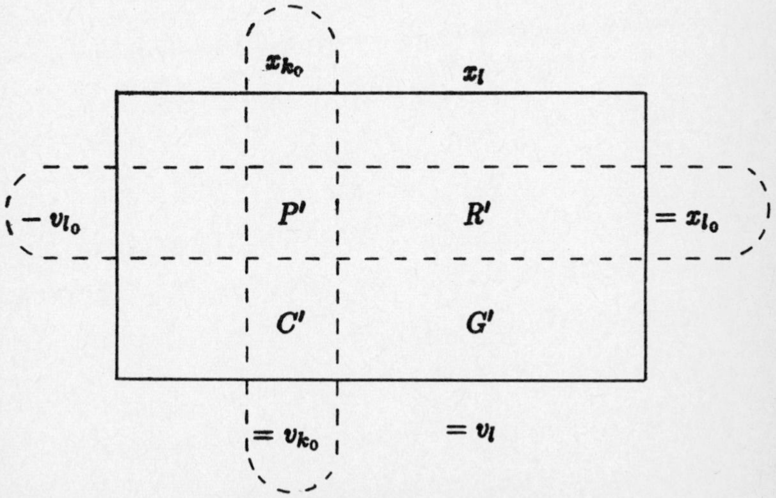

Figure 8.

can view C as the space of vectors orthogonal to the rows of the incidence matrix (one row for each node $i \in I$), and it follows then by elementary linear algebra that the complementary $D = C^{\perp}$ is the subspace spanned by these rows. Thus (taking $-u_i$ as the notation for the coefficient of the ith row in a general linear combination of the rows, for reasons apparent in a moment) we see that D consists of the vectors $v = (\ldots, v_j, \ldots)$ expressible in the form

$$v_j = - \sum_{i \in I} u_i e_{ij} \quad \text{for } j \in J \tag{13}$$

by some choice of numbers u_i. Referring to the definition of e_{ij}, we see further that (13) reduces to $v_j = u_{i_2} - u_{i_1}$, where i_1 is the initial node of the arc j and i_2 the terminal node.

The interpretation is this. The vector $u = (\ldots, u_i, \ldots) \in R^I$ is a *potential* on the nodes of the network, and $v = (\ldots, v_j, \ldots)$ is the corresponding vector of potential differences or *tensions*, the *differential* of u. Thus D is the tension or *differential space* of the network.

6. CONJUGATE COSTS AND MONOTONE RELATIONS.

The notion of a linear system of variables has been dualized, but that is not the only ingredient in a monotropic programming problem. We must also dualize the data embodied in the specification of a closed proper convex function f_j on R for each of the variables x_j, which includes the specification of the interval C_j associated with x_j (C_j being the effective domain of f_j). The machinery for this is already well developed in convex analysis. The natural dual of f_j is the closed proper convex function g_j on R *conjugate* to f_j. We take g_j as the cost function associated with the variable v_j dual to x_j, and its effective domain

$$D_j = \{v_j \mid g_j(v_j) \text{ finite } \}$$

as the interval associated with v_j.

The general formulas for passing between f_j and g_j are

$$\begin{aligned} g_j(v_j) &= \sup_{x_j \in R} \{v_j x_j - f_j(x_j)\} = \sup_{x_j \in C_j} \{v_j x_j - f_j(x_j)\}, \\ f_j(x_j) &= \sup_{v_j \in R} \{v_j x_j - g_j(v_j)\} = \sup_{v_j \in D_j} \{v_j x_j - g_j(v_j)\}. \end{aligned} \tag{14}$$

These formulas can often be used directly, but because we are dealing with convex functions of a single variable, there is an alternative method available for constructing g_j from f_j, or vice versa. This method, which involves inverting a generalized derivative relation and integrating, is often very effective and easy to carry out, and it yields other insights as well.

For every value of x_j the set

(15) $$\partial f_j(x_j) = \{v_j \in R \mid f_j(x_j+t) \geqq f_j(x_j) + v_j t \text{ for all } t \in R\}$$

consists of the "subgradients" of f_j at x_j in the general terminology of convex analysis, but in the one-dimensional case we are involved with here it is more appropriate to think of a "range of slopes" of f_j at x_j. This set is always a closed interval: in terms of the right derivative $f'_{j+}(x_j)$ and the left derivative $f'_{j-}(x_j)$ one has

(16) $$\partial f_j(x_j) = \{v_j \in R \mid f'_{j-}(x_j) \leqq v_j \leqq f'_{j+}(x_j)\}.$$

(For this to make sense even when $x_j \notin C_j$, the convention is adopted that $f'_{j-}(x_j)$ and $f'_{j+}(x_j)$ are both ∞ when x_j lies to the right of C_j but not $-\infty$ when x_j lies to the left of C_j; then f'_{j+} and f'_{j-} are nondecreasing functions on R.)

Technically one must view ∂f_j as a "multifunction" rather than a function, because $\partial f_j(x_j)$ can be empty or have more than one element. There is a remarkable function-like character, however, which becomes clear upon inspection of the graph set

(17) $$\Gamma_j = \text{gph } \partial f_j = \{(x_j, v_j) \in R \times R \mid v_j \in \partial f_j(x_j)\},$$

as illustrated in Figures 9 and 10.

In fact Γ_j is just like the graph of a nondecreasing function, but with "the vertical gaps filled in". More formally: there is a nondecreasing function ϕ_j in R (extended-real-valued) such that

(18) $$\Gamma_j = \{(x_j, v_j) \in R \times R \mid \phi_{j-}(x_j) \leqq v_j \leqq \phi_{j+}(x_j)\},$$

where $\phi_{j+}(x_j)$ is the right limit of ϕ_j at x_j and $\phi_{j-}(x_j)$ is the left limit. Indeed, any function ϕ_j chosen with

(19) $$f'_{j-}(x_j) \leqq \phi_j(x_j) \leqq f'_{j+}(x_j) \text{ for all } x_j.$$

has the property and yields

(20) $$f'_{j-}(x_j) = \phi_{j-}(x_j), \qquad f'_{j+}(x_j) = \phi_{j+}(x_j).$$

Moreover

(21) $$f_j(x_j) = f_j(\bar{x}_j) + \int_{\bar{x}_j}^{x_j} \phi_j(\xi)d\xi,$$

where $\bar{x}_j$ is any point in

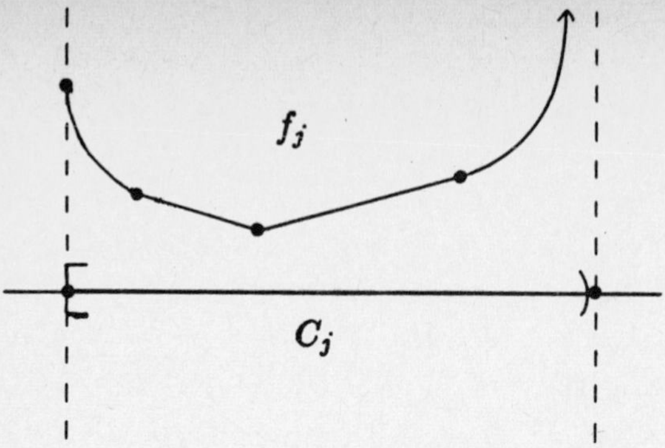

Figure 9.

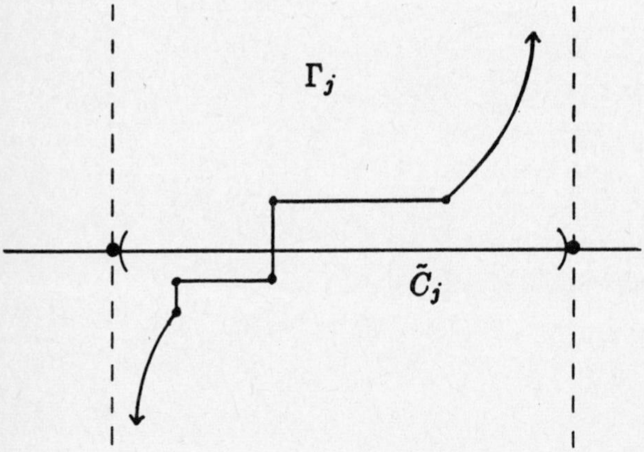

Figure 10.

$$\tilde{C}_j = \text{projection of } \Gamma_j \text{ on the horizontal axis} \tag{22}$$
$$= \{x_j \mid \partial f_j(x_j) \neq \emptyset\} \subset C_j .$$

In this sense, not only is Γ_j function-like, but f_j is uniquely determined up to an additive constant as the "integral" of Γ_j.

We note for later purposes that the set $\tilde{C}_j$ in (22) is a certain nonempty *interval* which includes the interior of C_j and therefore differs from C_j, if at all, only in a possible lack of one or the other endpoints. Figures 9 and 10 illustrate a case where $\tilde{C}_j$ does differ from C_j, and actually C_j is not closed, even though f_j is a closed proper convex function. In the great majority of applications C_j and $\tilde{C}_j$ coincide, and usually they are both closed too, but the possible discrepancy between these intervals cannot be ignored when it comes to stating theorems, cf. §8.

The sets Γ representable as augmented graphs of nondecreasing functions ϕ on R as in (16) (with ϕ not merely the constant function ∞ nor the constant function $-\infty$) are the so called *maximal monotone relations* in R × R. They are characterized by the property that their elements are totally ordered with respect to coordinatewise partial ordering of R × R, and no further elements can be added without this being violated. In terms of (16), (17), (18) and (19), then, every closed proper convex function f on R gives rise to a maximal monotone relation Γ in R × R, and every maximal monotone relation Γ arises in this way from some closed proper convex function f, which is unique up to an additive constant.

Piecewise linear functions f_j correspond to *staircase* Γ_j (comprised of finitely many line segments which are either vertical or horizontal), whereas piecewise quadratic functions f_j correspond to *polygonal* relations Γ_j (comprised of finitely many line segments, which do not have to be vertical or horizontal); see Figure 11.

The crucial fact for the purpose of constructing the conjugate g_j of f_j is that

$$v_j \in \partial f_j(x_j) \iff x_j \in \partial g_j(v_j),$$

or in other words

(23) $$\operatorname{gph} \partial g_j = \Gamma_j^{-1} = \{(v_j,x_j) \mid (x_j,v_j) \in \Gamma_j\};$$

see Figure 12.

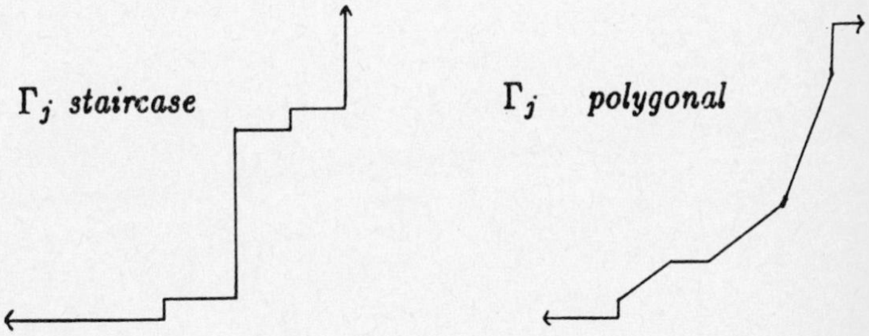

Figure 11.

Therefore one can determine g_j *by "differentiating"* f_j *to get* Γ_j, *passing to the "inverse" relation* Γ_j^{-1}, *and then "integrating"*. The appropriate constant of integration is fixed by the equation

$$g_j(\bar{v}_j) = \bar{v}_j\bar{x}_j - f_j(\bar{x}_j) \text{ for any } (\bar{x}_j,\bar{v}_j) \in \Gamma_j,$$

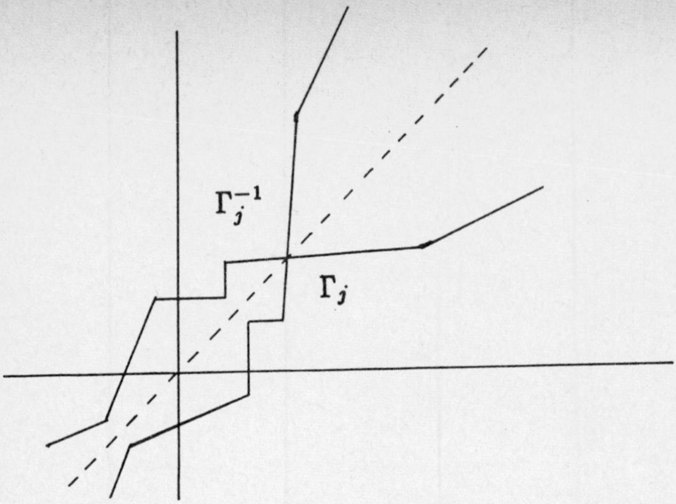

Figure 12.

which is a consequence of (14) and (15). Related to the finiteness interval D_j associated with g_j, as $\tilde{C}_j$ was to C_j, is the interval

$$\begin{aligned} \tilde{D}_j &= \text{projection of } \Gamma_j \text{ on the vertical axis} \\ &= \{v_j \in R \mid \partial g_j(v_j) \neq \emptyset\} \subset D_j. \end{aligned} \tag{24}$$

This always includes the interior of D_j.

It is obvious from this method of constructing conjugates that g_j is piecewise linear when f_j is piecewise linear, and g_j is piecewise quadratic when f_j is piecewise quadratic. Indeed, Γ_j^{-1} is "staircase" when Γ_j is "staircase", and Γ_j^{-1} is "polygonal". Furthermore the construction is quite easy to carry out in such cases. The conclusion to be drawn in the context of the next section will be that the dual of any piecewise linear or piecewise quadratic problem in monotropic programming can readily be written down.

7. DUAL PROBLEM AND EQUILIBRIUM PROBLEM.

In terms of a linear system of variables corresponding to a subspace C of R^J and a closed proper convex function f_j assigned to each $j \in J$ we have already introduced the *primal monotropic programming problem*

$$\text{(P)} \qquad \begin{aligned} &\text{minimize } \sum_{j\in J} f_j(x_j) =: F(x) \text{ over all} \\ &x = (\ldots,x_j,\ldots) \in C \text{ satisfying } x_j \in C_j \text{ for all } j \in J. \end{aligned}$$

We now introduce the *dual monotropic programming problem*

(D)
$$\text{maximize } -\sum_{j\in J} g_j(v_j) =: G(v) \text{ over all}$$
$$v = (\ldots,v_j,\ldots) \in D \text{ satisfying } v_j \in D_j \text{ for all } j \in J$$

and the *monotropic equilibrium problem*

(E)
$$\text{find } x = (\ldots,x_j,\ldots) \in C \text{ and } v = (\ldots,v_j,\ldots) \in D$$
$$\text{such that } (x_j,v_j) \in \Gamma_j \text{ for all } j \in J.$$

Here $D \in R^J$ gives the dual linear system as in §5, g_j is the cost function conjugate to f_j (with D_j its interval of finiteness), and Γ_j is the corresponding maximal monotone relation in $R \times R$ as in §6.

It is important to realize that because any member of the triple elements f_j, g_j, Γ_j, can be determined from any other, it is also true that any one of the problems (P), (D), and (E) generates the others. (The choice of constants of integration in passing from (E) to (P) and (D) makes no real difference, since it only affects the objective functions by a constant.) Applications do occur where (E) is paramount, as will be explained at the end of this section. The general connection between the problems is that (E) focuses on joint optimality conditions for the solutions to (P) and (D), or when viewed in the other direction, that (P) and (D) furnish variational principles for the solutions to (E).

Parallel to the expression of (P) in terms of a Tucker representation for the primal linear system as in (P') one has the expression of (D) as

(D')
$$\text{maximize } -\sum_{l\in L} g_l(v_l) - \sum_{k\in K} g_k(v_k)$$
$$\text{subject to } v_k \in D_k \text{ for all } k \in K,$$
$$-\sum_{k\in K} v_k a_{kl} = v_l \in D_l \text{ for all } l \in L.$$

Both problems can be viewed along with (E) in terms of a joint Tucker tableau for the primal and dual systems as in Figure 6. Obviously the general mathematical nature of the dual problem is the same as that of the primal problem, except for a change of sign in the objective (so that a concave function is maximized instead of a convex function minimized). The dual of the dual is the primal again; complete symmetry reigns.

It is hardly possible within the confines of this article to do more than hint at the wide range of situations encompassed by this paradigm. Some general observations that can be made on the basis of the preceding discussion are these. The dual of a piecewise linear problem is piecewise linear, and the dual of a piecewise quadratic problem is piecewise quadratic. In network programming, the dual of a problem involving flows is a problem involving potentials, and conversely.

A more specific illustration is the dual of the basic linear programming problem (4) in the case of upper bounds on the variables x_l, where

$$f_l(x_l) = c_l x_l + \delta(x_l | [0, d_l]) \text{ for } l \in L,$$

$$\text{(25)} \qquad f_k(x_k) = \begin{cases} \delta(x_k, [b_k, \infty)) & \text{for } k \in K_+, \\ \delta(x_k, [b_k, b_k]) & \text{for } k \in K_0. \end{cases}$$

(For brevity we omit the index set K_- at this time.) Simple calculations on the basis of (14) reveal that

$$g_l(v_l) = \begin{cases} 0 & \text{if } v_l \leqq c_l, \\ d_l(v_l - c_l) & \text{if } v_l > c_l, \end{cases}$$

$$\text{(26)}$$

$$g_k(v_k) = \begin{cases} b_k v_k + \delta(v_k | (-\infty, 0]) & \text{for } k \in K_+, \\ b_k v_k & \text{for } k \in K_0. \end{cases}$$

Thus in the corresponding dual (in form (D')) we maximize $-\Sigma_{k \in K} b_k v_k$ subject to $-v_k \geqq 0$, $-\Sigma_{k \in K} v_k a_{kl} = v_l \leqq c_l$, except that the constraint $v_l \leqq c_l$ can be violated to the tune of a *linear penalty* with cost coefficient d_l. Ordinary linear programming duality is included here as a special case: it corresponds to an "infinite penalty" for constraint violation. A change of variables $w_k = -v_k$ would reduce notation to the customary pattern of signs. Incidentally, the introduction of a linear penalty for violation of the constraints $x_k \geqq b_k$ in (P) would correspond dually to the introduction of an upper bound on the variable $w_k = -v_k$.

The nature of the relations Γ_l and Γ_k in this linear programming example is instructive too. These relations are displayed in Figure 13.

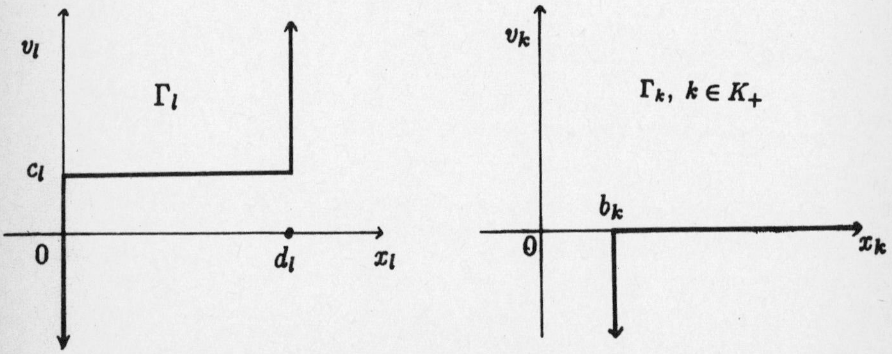

Figure 13.

For $k \in K_0$, Γ_k is simply the vertical line through the point b_k on the x_k-axis. Then the condition $(x_k,v_k) \in \Gamma_k$ reduces to the equation $x_k = b_k$ with no restriction placed on v_k, but the relations in Figure 13 express *complementary slackness* of various sorts. Thus for $k \in K_+$, for instance, the pairs $(x_k,v_k) \in \Gamma_k$ are the ones that satisfy

$$x_k \geqq b_k, \quad -v_k \geqq 0, \quad (x_k-b_k)(-v_k) = 0.$$

In network programming, on the other hand, the relations Γ_j can take quite a different sort of meaning. In this context x_j is the flow in the arc j, and v_j is the tension (potential difference). The condition $(x_j,v_j) \in \Gamma_j$ in the equilibrium problem (E) says that for the arc j only certain combinations of flow and tension are admissible. One is reminded of classical electrical networks and Ohm's Law, where $x_j = r_j v_j$ for a certain coefficient $r_j > 0$, the *resistance* of the arc j. Then Γ_j is a line in $R \times R$ with slope r_j.

Other classical electrical examples are those where the arc j represents an ideal *battery* (Γ_j a horizontal line through the point c_j on the v_j-axis, c_j being the potential difference across the terminals of the battery regardless of the current passing through it), an ideal *generator* (Γ_j a vertical line through the point b_j on the x_j-axis, b_j being the fixed current supplied by the generator regardless of the potential difference across its terminals), or an ideal *diode* (Γ_j the union of the nonnegative x_j-axis with the nonpositive v_j-axis). More general characteristic curves Γ_j can be obtained by imagining arc j that represent non-linear resistors or two-terminal "black boxes" with internal networks made up of various components such as have already been mentioned.

A fundamental problem in electrical networks, expressed in a mathematically polished, modern form, is the following. Given for each arc j of the network a "characteristic curve" Γ_j which is a *maximal monotone relation* in $R \times R$, find a circulation vector x and a differential vector v such that for every j the pair (x_j,v_j) lies on Γ_j. This is problem (E) for the linear systems of variables associated with the network.

Equilibrium problems in hydraulic networks arise similarly, and they also appear in the analysis of traffic. In economic networks (and general transportation problems) the potential at a node can often be interpreted as a commodity price at the node, and the tension v_j in the arc j is *price difference* across j. The relation Γ_j specifies the way in which the commodity flow in j responds to price difference. In Figure 13, for example, the relation Γ_1 says that when the price difference across 1 is greater than the cost per unit flow in 1 (namely c_1) the commodity flow should be at its upper bound d_1, whereas if the price difference is negative, the flow should be 0. If the price difference equals the unit transport cost, any amount of flow between 0 and d_1 is acceptable.

Often Γ_j can be interpreted in an economic model as a supply or demand relation. This leads to further instances of the equilibrium problem (E).

Before concluding this section, we mention another way that the standard duality scheme in linear programming can be viewed as a special case of monotropic programming duality. For this we imagine a Tucker tableau as in Figure 14 with two distinguished indices $\bar{k}$ and $\bar{l}$.

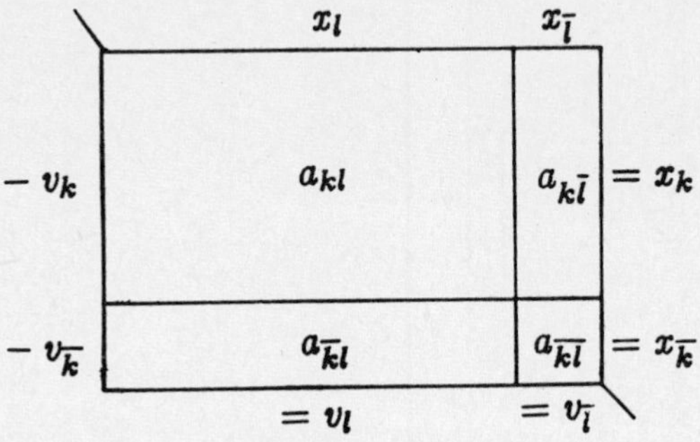

Figure 14.

We take

$$f_j(x_j) = \delta(x_j \mid [0,\infty)) = \begin{cases} 0 & \text{if } x_j \geqq 0 \\ \infty & \text{if } x_j < 0 \end{cases}$$

for every $j \in J = K \cup L$ except $\bar{k}$ and $\bar{l}$, but

$$f_{\bar{k}}(x_{\bar{k}}) \equiv x_{\bar{k}} \qquad (C_{\bar{k}} = (-\infty,\infty)),$$

$$f_{\bar{l}}(x_{\bar{l}}) = \delta(x_{\bar{l}} \mid [1,1]) = \begin{cases} 0 & \text{if } x_{\bar{l}} = 1, \\ \infty & \text{if } x_1 \neq 1. \end{cases}$$

The primal problem is then

$$(P_0) \qquad \begin{aligned} &\text{minimize } x_{\bar{k}} = \sum_{l \neq \bar{l}} a_{\bar{k}l} x_l + a_{\bar{k}\bar{l}} \text{ subject to} \\ &x_l \geqq 0 \text{ for } l \neq \bar{l},\ x_k = \sum_{l \neq \bar{l}} a_{kl} x_l + a_{k\bar{l}} \geqq 0 \text{ for } k \neq \bar{k}. \end{aligned}$$

One as

$$g_j(v_j) = \delta(v_j \mid (-\infty,0]) = \begin{cases} 0 & \text{if } -v_j \leqq 0 \\ \infty & \text{if } -v_j > 0 \end{cases}$$

for every j except $\bar{k}$ and $\bar{l}$, but

$$g_{\bar{l}}(v_{\bar{l}}) \equiv v_{\bar{l}} \qquad (D_{\bar{l}} = (-\infty,\infty))$$

$$g_{\bar{k}}(v_{\bar{k}}) = \delta(v_{\bar{k}}|[1,1]) = \begin{cases} 0 & \text{if } v_{\bar{k}} = 1, \\ \infty & \text{if } -v_{\bar{k}} \neq 1, \end{cases}$$

so the dual problem is

$$(D_0) \quad \begin{array}{l} \text{minimize} \quad -v_{\bar{1}} = \Sigma_{k \neq \bar{k}} v_k a_{k\bar{1}} + a_{\bar{k}\bar{1}} \text{ subject to} \\ -v_k \geqq 0 \text{ for } k \neq \bar{k}, \quad -v_1 = \Sigma_{k \neq \bar{k}} v_k a_{kl} + a_{kl} = 0 \text{ for } 1 \neq \bar{1}. \end{array}$$

For every j other than $\bar{k}$ and $\bar{1}$, the relation Γ_j is given by the union of the nonnegative x_j axis and the nonpositive v_j axis; it expresses "complementary slackness". The relation $\Gamma_{\bar{k}}$ is the horizontal line through at level 1 on the vertical scale, whereas $\Gamma_{\bar{1}}$ is the vertical line through level 1 on the horizontal scale.

8. THE MAIN THEOREMS OF MONOTROPIC PROGRAMMING.

The connection between problems (P), (D) and (E) is very tight. The theory is every bit as complete and constructive as in the familiar case of linear programming, but it applies to an enormously richer class of optimization models.

Duality Theorem. *If any of the following conditions is satisfied,*

(a) the primal problem (P) *has feasible solutions and finite optimal value* inf(P), *or*

(b) the dual problem (D) *has feasible solutions and finite optimal value* sup(D), *or*

(c) the primal problem (P) *and the dual problem* (D) *both have feasible solutions,*

then all three hold, and

$$\inf(P) = \sup(D).$$

Equilibrium Theorem. *A pair* (x,v) *solves the equilibrium problem* (E) *if and only if* x *solves the primal problem* (P) *and* v *solves the dual problem* (D).

Results on the existence of solutions to (P), (D) and (E) require a distinction between the intervals C_j and $\tilde{C}_j$, and between D_j and $\tilde{D}_j$, where $\tilde{C}_j$ and $\tilde{D}_j$ are given by (22) and (24). Whereas a *feasible* solution to (P) is an $x \in C$ such that $x_j \in C_j$ for all $j \in J$, a *regularly feasible* solution is defined to be an $x \in C$ such that $x_j \in \tilde{C}_j$ for all $j \in J$. Likewise a *regularly feasible* solution to (D) is defined to be a $v \in D$ such that $v_j \in \tilde{D}_j$. This distinction falls away in the case of so-called *regular* monotropic programming problems, where $C_j = \tilde{C}_j$ and $D_j = \tilde{D}_j$ for all $j \in J$ (as is true in particular when $\tilde{C}_j$ and $\tilde{D}_j$ are closed, i.e. when Γ_j projects horizontally and vertically onto closed intervals). Important to keep in mind as regular problems in this sense are the problems of piecewise linear programming or piecewise quadratic programming.

Existence Theorem.

(a) The primal problem (P) *has an optimal solution if and only if* (P) *has a feasible solution and* (D) *has a regularly feasible solution.*

(b) The dual problem (D) *has an optimal solution if and only if* (D) *has a feasible solution and* (P) *has a regular feasible solution.*

(c) The equilibrium problem (E) *has a solution if and only if* (P) *and* (D) *both have regularly feasible solutions.*

Corollary. *In the case of regular problems* (P) *or* (D), *an optimal solution exists if a feasible solution exists and the optimal value is finite.*

By combining the existence theorem with the equilibrium theorem, we obtain the following characterization of optimal solutions to (P) and (D) that is the basis of most computational procedures in monotropic programming.

Optimality Theorem.

(a) Suppose that the primal problem (P) *has at least one regularly feasible solution. Then for* x *to be an optimal solution to* (P) *it is necessary and sufficient that* x *be regularly feasible and such that the dual linear system has a vector satisfying:*

$$v_j \in \partial f_j(x_j) = [f'_{j-}(x_j), f'_{j+}(x_j)] \text{ for all } j \in J$$

(Such a v *is an optimal solution to* (D).*)*

(b) Suppose that the dual problem (P) *has at least one regularly feasible solution. Then for* v *to be an optimal solution to* (D) *it is necessary and sufficient that* v *be regularly feasible and such that the primal linear system has a vector* x *satisfying*

$$x_j \in \partial g_j(v_j) = [g'_{j-}(v_j), g'_{j+}(v_j)] \text{ for all } j \in J$$

(Such an x *is an optimal solution to* (P).*).*

9. SOLUTION BY PIVOTING METHODS.

An important feature of monotropic programming problems is the possibility of solving them by techniques based on repeated pivoting transformations of Tucker tableaus for the underlying linear systems. Such techniques can in some cases be viewed as generalizations of the various forms of the simplex method in linear programming, but they may also be based on distinctly different approaches involving phenomena of another order.

Our aim is to exploit the fact that the primal and dual linear system of variables can be represented in more than one way by tableaus of the kind in Figure 6. Even though only one such tableau may have been given to us initially, others can be generated by pivoting. We wish to make use of their special properties as a means of constructing a sequence of feasible solutions to (P) or (D) that converges to, or in finitely many steps actually reaches, an optimal solution.

The optimality test provided by the last theorem in the preceding section has a central role in this context. In terms of a Tucker tableau associated with a partition of J into index sets K and L, it says the following: *a regularly feasible solution* x *to* (P) *is optimal if and only if for some choice of values* v_k *in the intervals* $\partial f_k(x_k)$, $k \in K$, *the corresponding values* $v_l = -\Sigma_{k\in K} v_k a_{kl}$ *satisfy* $v_l \in \partial f_l(x_l)$ *for all* $l \in L$. Likewise in the dual problem, *a regularly feasible solution* v *to* (D) *is optimal if and only if for some choice of values* $x_l \in \partial g_l(v_l)$ $l \in L$, *the corresponding values* $x_k = \Sigma_{l\in L} a_{kl} x_l$ *satisfy* $x_k \in \partial g_k(v_k)$ *for all* $k \in K$. Our attention actually is directed at the negatives of these conditions, which, as it turns out, can be stated in much sharper form than might be expected. This sharper form is based on the next theorem, which we present in terms of general intervals D_j' and C_j', not just $\partial f_j(x_j)$ and $\partial g_j(v_j)$, because of its other applications.

Feasibility Theorem.

(a) Let C_j' *denote a nonempty real interval (not necessarily closed) for each* $j \in J$. *If there does not exist an* $x \in C$ *satisfying* $x_j \in C_j'$ *for every* $j \in J$, *then in fact there exists a Tucker representation (with* J *partitioned into some* K *and* L*) and an index* $k_0 \in K$ *such that for no choice of values* $x_l \in C_l'$ *for* $l \in L$ *does the number* $x_{k_0} = \Sigma_{l\in L} a_{k_0 l} x_l$ *satisfy* $x_{k_0} \in C_{k_0}'$

(b) Let D_j' *denote a nonempty real interval (not necesarily closed) for each* $j \in J$. *If there does not exist a* $v \in D$ *satisfying* $v_j \in D_j'$ *for every* $j \in J$, *then in fact there exists a Tucker representation (with* J *partitioned into some* K *and* L*) and an index* $l_0 \in L$ *such that for no choice of values* $v_k \in C_k'$ *for* $k \in K$ *does the number* $v_{l_0} = -\Sigma_{k\in K} v_k a_{kl_0}$ *satisfy* $v_{l_0} \in C_{l_0}'$.

This result is valid in particular as a test of feasibility in (P) and (D) (the case of $C_j' = C_j$, $D_j' = D_j$) and regular feasibility ($C_j' = \tilde{C}_j$, $D_j' = \tilde{D}_j$). Our concern at present, though, is with the case of $C_j' = \partial g_j(v_j)$ and $D_j' = \partial f_j(x_j)$, where (a) and (b) characterize nonoptimality of v in (D) and of x in (P), respectively. Pivoting rules do exist for producing special Tucker tableaus and indices k_0 or l_0 with the properties of the theorem. Rather than explain such rules here, which would take too much space, we shall try to indicate the way that the conditions provided by the theorem can be used in optimization.

The basic idea we need to work with is that of trying to improve a given feasible solution x to (P) by changing only one "independent" variable at a time the primal independent variables relative to a particular Tucker tableau being the ones indexed by the set L in the partition $J = (K|L)$. Consider the situation in Figure 15, where a certain index $l_0 \in L$ has been singled out (first diagram).

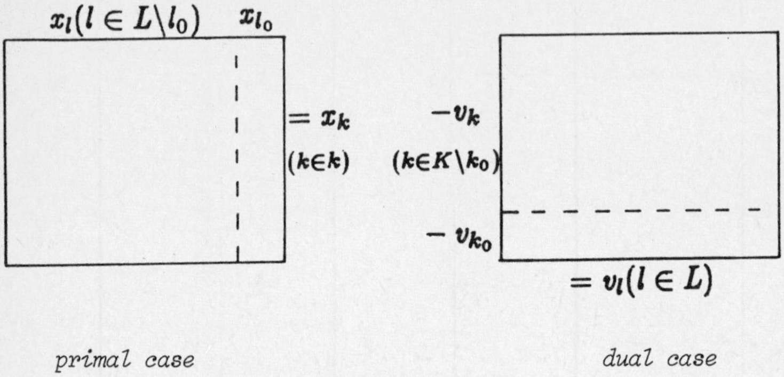

Figure 15.

We want to look at feasible modifications of x that leave fixed all the values x_l for $l \neq l_0$. Such a modification depends only on specifying the value of a single parameter $t \in R$:

$$x'_{l_0} = x_{l_0} + t, \qquad x'_l = x_l \text{ for all } l \in L\backslash l_0,$$

$$x'_k = \sum_{l \in L} a_{kl} x'_l = x_k + a_{kl_0} t \text{ for all } k \in K.$$

The corresponding objective value is

$$\phi(t) = F(x') = \sum_{l \neq l_0} f_l(x_l) + f_{l_0}(x_{l_0} + t) + \sum_{k \in K} f_k(x_k + a_{kl_0} t)$$

where $\phi(0) = F(x)$ is the objective value already at hand. We are interested in the existence of a t such that $\phi(t) < \phi(0)$, because x' is than a "better" feasible solution to (P) than x. (It is still feasible because $F(x') < F(x) < \infty$ implies $x'_j \in C_j$.) When such a t exists, we shall say that *monotropic improvement* of x is possible in (P) which respect to the Tucker representation in question and the index $l_0 \in L$. The exact choice of the stepsize t, whether by complete minimization of ϕ or some other device, need not concern us here.

The important thing is that the condition in part (b) of the Feasibility Theorem in the case of the intervals $D'_j = \partial f_j(x_j)$ holds for a particular Tucker representation and index $l_0 \in L$ if and only if monotropic improvement of x is possible for this Tucker representation and index. This can be verified by a calculation of the right and left derivatives of the convex function ϕ at 0. A similar result holds for the dual problem, where one is interested in improving a given regularly feasible solution v without changing the values v_k for $k \neq k_0$

in a certain tableau (see Figure 15 again, second diagram).

The situation is summarized in the next result.

Improvement Theorem.

(a) If x *is a regularly feasible solution to* (P) *which is not optimal, then monotropic improvement of* x *is possible with respect to some Tucker representation and index* $l_0 \in L$.

(b) If v *is a regularly feasible solution to* (D) *which is not optimal, then monotropic improvement is possible with respect to some Tucker representation and index* $k_0 \in K$.

This theorem leads to optimization procedures which alternate between pivoting routines that construct special Tucker tableaus and line searchers that minimize special convex functions ϕ. In network programming, of course, where the Tucker tableaus correspond to spanning trees, the pivoting can be carried out in a combinatorial manner. The monotropic improvement steps then amount to modifying a circulation x only around one closed path at a time, and modifying a differential v only across one cut at a time.

10. GENERALIZED SIMPLEX METHODS.

A nice illustration of these ideas, although by no means the only one, is the way that the simplex method of linear programming can be extended to monotropic programming problems in general. Let us call a value of x_j a *breakpoint* of f_j if $f'_{j-}(x_j) \neq f'_{j+}(x_j)$, i.e. if Γ_j has a vertical segment at x_j. (A finite endpoint of C_j that belongs to C_j is a breakpoint in this sense, in particular.) Let us say further that a regularly feasible solution x to (P) is *nondegenerate* if there is a Tucker tableau with partition $J = (K|L)$ such that x_k is not a breakpoint of f_k for any $k \in K$.

Such a tableau, if one exists, is very easy to construct by pivoting: simply exchange breakpoint indices with nonbreakpoint indices until all the breakpoint indices correspond to columns of the tableau instead of rows. One then has a quick test of the optimality of x. The intervals $\partial f_k(x_k)$ for $k \in K$ all consist of just a single point, namely the derivative value $f'_k(x_k)$; taking this as v_k and defining $v_l = -\Sigma_{k \in K} v_l a_{kl}$, check whether $v_l \in \partial f_l(x_l)$ for every $l \in L$. If "yes", then x is an optimal solution to (P) (and v is an optimal solution to (D); cf. the optimality theorem in §8). If "no", then for some index l_0 one has $v_{l_0} \notin \partial f_{l_0}(x_{l_0})$. Such an index l_0 satisfies the condition in the feasibility theorem in §9 for the intervals $C'_j = \partial f_j(x_j)$, and it therefore signals the possibility of monotropic improvement of x. This improvement can be carried out, and x is replaced by x'. If x' is again nondegenerate, the tableau can be restored to proper form with respect to x' if necessary, and the optimality test repeated.

For problems in piecewise linear programming, some refinements of this general simplex procedure are possible. Let us say that x is *quasi-extreme* for (P) if there is a Tucker tableau with partition J = (K|L) such that x_l is a breakpoint of f_l for every $l \in L$. Inasmuch as there are only finitely many breakpoints for each of the cost functions in a piecewise linear programming problem, and also only finitely many possible Tucker tableaus, there can be only finitely many feasible solutions that are quasi-extreme. If the procedure already described is initiated with such a feasible solution x, and if the line search in each monotropic improvement step is carried out with exactitude (which is easy because of piecewise linearity), then all the successive feasible solutions generated by the procedure will be quasi-extreme. Under the *nondegeneracy assumption* that every quasi-extreme feasible solution to (P) is nondegenerate, the procedure can be continued until, after finitely many steps, optimality is achieved.

This general version of the simplex method reduces to the classical method when applied to a linear programming problem in standard form. It also includes the modified simplex method for linear programming problems with upper bounds. It can be used directly on problems obtained from linear programming problems by penalty representation of constraints.

A dual version of all this can be written down in terms of problem (D), of course.

REFERENCES

[1] R.T. Rockafellar, Network Flows and Monotropic Optimization, Wiley-Interscience, 1984.

GENERALIZED DIFFERENTIABILITY, DUALITY AND OPTIMIZATION FOR PROBLEMS DEALING WITH DIFFERENCES OF CONVEX FUNCTIONS

J.-B. HIRIART-URRUTY

ABSTRACT. *A function is called d.c. if it can be expressed as a difference of two convex functions. In the present paper we survey the main known results about such functions from the viewpoint of Analysis and Optimization.*

INTRODUCTION

The analysis and optimization of convex functions have received a great deal of attention during the last two decades. If we had to choose two key-words from these developments, we would retain the concept of *subdifferential* and the *duality theory*. As it usual in the development of mathematical theories, people had since tried to extend the known definitions and properties to new classes of functions, including the convex ones. For what concerns the generalization of the notion of subdifferential, tremendous achievements have been carried out in the past decade and any mathematician who is faced with a nondifferentiable nonconvex function has now a panoply of generalized subdifferentials or derivatives at his disposal. A lot remains to be done in this area, especially concerning *vector-valued* functions ; however we think the golden age for these researches is behind us.

Duality theory has also fascinated many mathematicians since the underlying mathematical framework has been laid down in the context of Convex Analysis. The various duality schemes which have emerged in the recent years, despite of their mathematical elegance, have not always proved as powerful as expected.

The present paper is of a limited scope since it deals with generalized differentiability, duality and optimization for problems dealing with *differences of convex functions*. A real-valued function f defined on a convex set X is called *d.c.* on X (abbreviation for difference of convex functions on X) if there are two convex functions g and h such that :

$$\forall x \in X \quad f(x) = g(x)-h(x). \tag{0.1}$$

Why to study such functions ? At first sight it might look as a mathematician's crotchet. There are in fact many reasons for considering such functions. Firstly there are some "mathematical" reasons for doing so. The class of d.c. functions on X, denoted by DC(X), is clearly the vector space generated by the cone of convex functions on X (denoted Conv(X) throughout). It also happens that some classes $K(X)$ of functions recently considered in the literature, and including for example convex or C^2 functions on an open convex set X, give rise to the vector space DC(X). The framework is typically as follows :

$$\begin{aligned} &K(X) \text{ is a convex cone of functions on } X, \\ &\text{Conv }(X) \subset K(X) \subset DC(X), \qquad\qquad (0.2)\\ &K(X) - K(X) = DC(X). \end{aligned}$$

Another mathematical reason explaining our interest in d.c. functions is that DC(X) is dense in the set C(X) of continuous functions over a compact convex set X, endowed with the topology of uniform convergence over X. This is a mere application of the STONE-WEIERSTRASS theorem. Density results of the same vein can be derived when X is no more closed or bounded. Finally the class DC(X) enjoys a remarkable stability with respect to operations usually encountered in Optimization, like taking the maximum (or minimum) of a finite number of functions, taking the product (or the quotient, the sum, etc...) of functions. We shall harck back to it later on.

Further incentives for studying d.c. functions come from applications. Whean dealing with a nonconvex optimization problem, it often happens that the data are actually d.c. (sometimes after a transformation of the original problem or through some dualization scheme). Our genuine feeling is that most of the "reasonable" optimization problems actually involve d.c. functions, even if we are not always able to recognize them as such ! To begin with, let us mention very simple examples.

Example 0.1. Let A be any (n,n) symmetric matrix and let Q be the associated quadratic form on $\mathbb{R}^n$, namely :

$$Q(x) = \frac{1}{2} < Ax,x >.$$

Q is obviously d.c. on $\mathbb{R}^n$ and there are several ways of finding positive semi-definite A^+ and A^- such that :

$$Q(x) = \frac{1}{2} < A^+x,x > - \frac{1}{2} < A^-x,x>. \qquad (0.3)$$

Example 0.2. Let S be an arbitrary nonempty subset of $\mathbb{R}^n$. It turns out that the square of the distance to S, denoted by d_S^2, is d.c. :

$$\forall x \in \mathbb{R}^n \quad d_S^2(x) = \|x\|^2 - \{ \|x\|^2 - d_S^2(x)\}. \qquad (0.4)$$

Surprisingly enough, $h : x \to h(x) = \frac{1}{2}\|x\|^2 - \frac{1}{2} d_S^2(x)$ is convex whatever S be (ASPLUND, 1973).d_S^2 is thus always d.c. ; it is convex whenever S is convex.

Example 0.3. Let A be a symmetric positive definite (n,n) matrix and let λ_M be the largest eigenvalue of A. By transforming RAYLEIGH's formulation :

$$\frac{\lambda_M}{2} = \max \{\frac{1}{2} < Ax, x > ; \ \|x\| \leq 1\}$$

via dualization schemes which will be mentioned later (section IV), we obtain that :

$$-\frac{\lambda_M}{2} = \min_{x^* \in \mathbb{R}^n} \{\|x^*\| - \frac{1}{2} < A^{-1} x^*, x^* >\} \qquad (0.5)$$

or

$$-\frac{\lambda_M}{2} = \min_{x^* \in \mathbb{R}^n} \{\frac{1}{2} \|x^*\|^2 - \sqrt{< Ax^*, x^* >}\} . \qquad (0.6)$$

Calculating λ_M can therefore be viewed as a problem of minimizing (or maximizing) a d.c. function.

Looking at the above-displayed examples or considering more gèneral d.c. functions give rise to the following fundamental questions.

Q1. *How to recognize d.c. functions ?*

In view of the function f itself, or the derivative of f (or some generalized version of it), or the second derivative of f (in whatever sense), how to decide that f is actually d.c. ?

Q2. *Given a d.c. function, what is the "best" decomposition of it as a difference of convex functions ?*

There are infinitely many ways of decomposing a d.c. function as differences of convex functions. What should it mean that a decomposition is "better" than another one ?

Q3. *What (more !) about optimality conditions for problems dealing with d.c. functions ?*

D.c. functions are examples of locally Lipschitz functions for which a great deal has been done concerning optimality conditions. So, one may wonder what more can be said when the involved functions are d.c.. A related but more fundamental question concerns the duality schemes : *what is the involution $f = g - h \to \varphi(g-h)$ corresponding to $f \to \varphi = f^*$ for convex* *f* ? We shall see in that respect that TOLAND's involution $g-h \to h^* - g^*$ takes root in a basic formula yielding the conjugate (in the sense of Convex Analysis) of g-h.

Q4. *How to use the richness of structure of d.c. functions for designing global minimization algorithms ?*

Since convexity is present twice in the decomposition of a d.c. function (in an antagonistic way however), one may think of using this particular structure to devise algorithms which would converge to a global minimum (or a global maximum) of the given function.

We do not pretend to answer fully all these questions here. Our intention in this paper is to take into account some recent contributions in this area of research and to pose in a clear-cut manner the problems which remain unsolved. Before going further we lay out the setting of our presentation. Throughout the underlying space will be $\mathbb{R}^n$ and, for the sake of simplicity, we suppose the convex set X on which are defined d.c. functions is the whole space, so that we are concerned with DC ($\mathbb{R}^n$). Most of the conclusions derived throughout extend to the DC(X), where X is an open convex set. To assume a finite-dimensional setting is a stronger limitation. Solving some partial differential equations amount to solving $f'(u) = 0$ where f is a d.c. function defined on an appropriate Hilbert space. These approaches however go beyond the scope of this paper. The paper is organized as follows :

I. D.c. functions : first properties ;

II. Recognizing a d.c. function ;

III. Decomposing d.c. functions ;

IV. Optimality conditions, duality for d.c. functions ;

V. Preview on minimization procedures for d.c. functions.

I. D.C. FUNCTIONS : FIRST PROPERTIES

Let $f \in DC(\mathbb{R}^n)$, so that there exist convex functions g and h from $\mathbb{R}^n$ into $\mathbb{R}$ such that $f = g-h$. Some properties of f are directly inherited from those of convex functions ; we list the main of them.

First of all, f is *locally Lipschitz* on $\mathbb{R}^n$. The derivative of f(*) does exist almost everywhere (a.e.) and if one denotes by Ω_f the set of points where $\nabla f(x)$ fails to exist, one clearly has that $\Omega_f \subset \Omega_g \cup \Omega_h$ whatever the decomposition of f as a difference of convex g and h. Since the nature of the set of points where a convex function is not differentiable is known in a great detail, one can therefore get a better insight into the structure of Ω_f. The directional derivative $d \to f'(x;d)$ exists everywhere and :

$$f'(x;d) = g'(x;d) - h'(x;d) \text{ for all } x \text{ and } d. \qquad (1.1)$$

So, "the tangent problem at x" defined by $f'(x;\cdot)$ is itself d.c. The functions for which $f'(x;\cdot)$ can be written as a difference of two positively homogeneous (finite) convex functions are called *quasi-differentiable* by DEMYANOV and his associates (cf. [12], [13] for example). For various reasons we prefer the terminology "*tangentially d.c.*" to "quasi-differentiable" for such functions. The gap between the class of d.c. functions and that of tangentially d.c. functions is actually small. As for example, f is tangentially d.c. whenever it is differentiable, while a little more is required on ∇f for f to be d.c. (see next section).

Concerning the generalized gradient (in CLARKE's sense) of f, the following estimate holds :

$$\partial f(x) \subset \partial g(x) - \partial h(x) \text{ for all } x. \qquad (1.2)$$

This estimate however may be very coarse. In fact, by writting f as $(g+\varphi) - (h+\varphi)$ with an ad hoc convex function φ, the estimate (1.2) can be rendered as rough as desired ! An exact evaluation of $\partial f(x)$ would require us to know the generalized derivative of the $\mathbb{R}^2$-valued function $(g,-h)^T$. But, for that purpose, to know that g and h are convex do not help very

---

(*) Derivatives in the sense of GATEAUX, HADAMARD or FRECHET are equivalent for $f \in DC(\mathbb{R}^n)$.

much. There is however a consequence one can draw immediately from (1.2) and which is as follows : *$\partial f(x)$ is reduced to $\{\nabla f(x)\}$ a.e.*. Surprisingly enough and contrary to what happens for convex functions, *$\partial f(x_0)$ does not necessarily reduce to a singleton when f is differentiable at x_0.*

Example 1.1. (from [45]).

Lef $f : \mathbb{R}^2 \to \mathbb{R}$ be defined as $f(\xi_1, \xi_2) = |\xi_2 - \xi_1^3| - |\xi_2|$. For reasons which will appear later (section II) f is d.c. One verifies that f is differentiable at (0,0) with $\nabla f(0,0) = (0,0)$, while $\partial f(0,0) = \{0\} \times [-2,2]$. Calculating $\partial f(\xi_1, \xi_2)$ at points (ξ_1, ξ_2) satisfying $\xi_2 = \xi_1^3$ or $\xi_2 = 0$ gives a good idea of how $\partial f(\xi_1, \xi_2)$ behaves when (ξ_1, ξ_2) approaches (0,0).

Therefore there may be points where f is differentiable without being strictly differentiable. Let Ω_f^1 denote the set (of null measure) where f fails to be strictly differentiable. If $x_0 \in \Omega_f^1 \setminus \Omega_f$, it comes from (1.1) that :

$$\partial g(x_0) = \partial h(x_0) + \nabla f(x_0), \tag{1.3}$$

that is $\partial g(x_0)$ is a translation of $\partial h(x_0)$. For the f displayed in Example 1.1, a decomposition as a difference of convex functions is as follows :

$$f(\xi_1, \xi_2) = 2 \max \{(\xi_1^3)^+, \xi_2 + (\xi_1^3)^-\} - (|\xi_1|^3 + 2\xi_2^+).$$

$$= g(\xi_1, \xi_2) - h(\xi_1, \xi_2).$$

At $x_0 = (0,0)$ we have that $\partial g(x_0) = \partial h(x_0) = [0,2]$. The phenomenon we have observed raises the question whether the generalized gradient is appropriate to d.c. functions ; for further discussion about the various first-order generalized derivatives, the reader may refer to [25]. Given any $\varphi \in \mathrm{Conv}(\mathbb{R}^n)$ and $a, b \in \mathbb{R}^n$, the difference $\varphi(b) - \varphi(a)$ can be represented in an integral form as follows :

$$\varphi(b) - \varphi(a) = \int_a^b < \partial\varphi(a + t(b-a)), b-a > dt, \tag{1.4}$$

where the right-hand means the integral of the multifunction $\Gamma_{a,b}$: $t \rightrightarrows < \partial\varphi(a+t(b-a)), b-a >$ over [a,b]. This is a formal writing since the integrals of all the measurable selections of $\Gamma_{a,b}$ over [a,b] yield the same value, namely $\varphi(b) - \varphi(a)$. Despite of the inclusion (1.2) and the possible discrepancy between the $\partial f(x)$ and $\{\nabla f(x)\}$, an integral representation analogous to (1.4) holds true for a d.c. function f. For any locally Lipschitz function f on $\mathbb{R}^n$ we know that :

$$f(b) - f(a) \in \int_a^b < \partial f(a+t(b-a)), b-a > dt. \qquad (1.5)$$

Now, (1.4) holds for $g(b)-g(a)$ and $h(b)-h(a)$ whenever f is expressed as a difference of convex functions g and h. Since the integral of the sum of two integrable multifunctions is the sum of their integrals, we have :

$$f(b)-f(a) = \int_a^b < \partial g(a+t(b-a)) - \partial h(a+t(b-a)), b-a > dt.$$

Whence

$$f(b)-f(a) = \int_a^b < \partial f(a+t(b-a)), b-a > dt. \qquad (1.6)$$

Knowing ∂f therefore allows us to recover f.

Concerning second-order derivatives, d.c. functions inherit from convex functions they are "*twice-differentiable a.e.*". Let us make this statement more precise. The multifunction ∂f is said to be differentiable at x_o if f is differentiable at x_o and if there exists a linear mapping denoted by $\nabla^2 f(x_o)$ such that :

$$\|\partial f(x) - \nabla f(x_o) - \nabla^2 f(x_o)(x-x_o)\| = o(\|x-x_o\|), \qquad (1.7)$$

or in other words :

$$\forall \eta > 0, \exists \delta > 0, \forall x \text{ with } \|x-x_o\| \leq \delta, \forall x^* \in \partial f(x),$$

$$\|x^* - \nabla f(x_o) - \nabla^2 f(x_o)(x-x_o)\| \leq \eta \|x-x_o\|.$$

$\nabla^2 f(x_o)$ is then a symmetric mapping which is called the derivative of ∂f at x_o. As an application of MIGNOT's differentiability theorem on maximal monotone multifunctions [34, Theorem 1.3] we have that the subdifferential multifunction of $\varphi \in \mathrm{Conv}(\mathbb{R}^n)$ is differentiable a.e. on $\mathbb{R}^n$. It is therefore just a corollary to claim that "*the generalized gradient of a d.c. function is differentiable a.e.*". Another formulation, better known in the context of Convex Analysis, states that a convex function on $\mathbb{R}^n$ has a second-order TAYLOR expansion at almost all points of $\mathbb{R}^n$ (ALEXANDROFF, 1939).

That means that, at almost every $x_0 \in \mathbb{R}^n$, there is a (symmetric) linear mapping denoted by $A^2f(x_0)$ such that :

$$f(x) = f(x_0) + \langle \nabla f(x_0), x-x_0 \rangle + \frac{1}{2} \langle A^2 f(x_0)(x-x_0), x-x_0 \rangle + o(\|x-x_0\|^2). \qquad (1.8)$$

Clearly f has a second-order expansion at x_0 whenever ∂f is differentiable at x_0. Thus "*a d.c. function has a second-order TAYLOR expansion a.e.*".

Let Ω_f^2 denote the set of null measure where ∂f fails to be differentiable. It readily comes from (1.7) that $\partial f(x_0) = \{\nabla f(x_0)\}$ whenever ∂f is differentiable at x_0. We thus summarize the differentiability properties of a d.c. function f by stating :

$$\begin{aligned} &\Omega_f \subset \Omega_f^1 \subset \Omega_f^2 \\ &\Omega_f^2 \text{ is of null measure.} \end{aligned} \qquad (1.9)$$

There are limitations in extending properties of convex functions to d.c. functions. We briefly mention here some of them.

The graph of the subdifferential of a convex function on $\mathbb{R}^n$ is of a very special structure since it is a *Lipschitz manifold* of $\mathbb{R}^n \times \mathbb{R}^n$ (see [43] for a recent account on that subject). The graph of the generalized gradient of a d.c. function is no more a Lipschitz manifold.

Example 1.2. Let $f : \mathbb{R} \to \mathbb{R}$ be defined by $f(0) = 0$ and $f'(x) = |x|^{1/2}$ for all x. $f \in DC(\mathbb{R})$ but the graph of f' is not a Lipschitz curve in $\mathbb{R}^2$.

Apart from some nasty points like 0 in the previous example, the graph of the generalized gradient of a d.c. function looks pretty much alike a Lipschitz manifold. So, most of the geometrical properties of tangent cones to the graph of the subdifferential of a convex function (such as displayed in [43]) should have their counterparts for d.c. functions. Another property of convex functions which is no longer true for d.c. functions is the following : *a limit of d.c. functions is not necessarily d.c.*.

Example 1.3. (from [45])

Let $f_n : \mathbb{R} \to \mathbb{R}$ be defined as $f_n(x) = \text{Min}\,\{|x - \frac{1}{k}|\,,\ k = 1,\dots,n\}$. f_n is d.c. on $\mathbb{R}$ but the derivative of $f = \lim_{n\to+\infty} f_n$ is not of bounded variation in a neighbourhood of 0, i.e., f is not d.c. around 0. The same example could serve to show that the infimum (or the supremum) of an infinite family of d.c. functions is not necessarily d.c..

A more severe drawback is that, contrary to convex functions, a function $f : \mathbb{R}^n \to \mathbb{R}$ may be d.c. on lines (i.e., $t \to f(a + t(b-a))$ is d.c. whatever a and b) without being d.c. on the whole space $\mathbb{R}^n$. YOMDIN has shown us examples of such functions. This state of affairs is somewhat baffling since, as recalled in the next section, d.c. functions on the real line bear an easy characterization. All the properties of d.c. functions displayed in this section can serve in their negative form : a function which does not satisfy one of the properties mentioned is not d.c..

Example 1.4. Let A be a Borel set of $\mathbb{R}$ satisfying the next property : for all nonempty open interval I of $\mathbb{R}$, $\lambda(A \cap I) > 0$ and $\lambda(A^c \cap I) > 0$. The function f defined on $\mathbb{R}$ by :

$$f(x) = \int_0^x 1_A(t)\,dt$$

is strictly increasing and locally Lipschitz on $\mathbb{R}$. Is f d.c. on $\mathbb{R}$? Although there are various reasons for answering "no", we retain the following one : $\partial f(x) = [0,1]$ for all $x \in \mathbb{R}$ so that $\partial f(x)$ does not reduce to $\{f'(x)\}$ a.e..

II. RECOGNIZING D.C. FUNCTIONS

Whether a locally Lipschitz function *on* $\mathbb{R}$ is d.c. or not depends on the variation of f' (defined a.e.). f is d.c. on $\mathbb{R}$ if and only if f' is of bounded variation on compact intervals of $\mathbb{R}$. This is easy to imagine since, according to JORDAN's decomposition, f' is of bounded variation if and only if it can be expressed as a difference of two increasing functions. A further property which is worth noticing is the following.

THEOREM 2.1. *Let f be a differentiable d.c. function on $\mathbb{R}$. f is then continuously differentiable and can be written as a difference of (continuously) differentiable convex functions.*

Proof. f', as a derivative, satisfies DARBOUX' property, that is : the image by f' of any interval is an interval. Therefore, f' is continuous whenever it is of bounded variation. Hence f' can be written as a difference of two continuous increasing functions.□

To recognize d.c. functions among locally Lipschitz functions on $\mathbb{R}^n$ is not an easy matter. We will give some indications about that, according to what is at our disposal : the function itself, or its first or second (generalized) derivative.

II.1. Apart from some contributions peculiar to the two-dimensional case ([58] for example), the main result on recognizing d.c. functions on $\mathbb{R}^n$ by referring only to the function goes back to HARTMAN (1959). Before stating it, we recall that $f : \mathbb{R}^n \to \mathbb{R}$ is said *locally d.c.* on $\mathbb{R}^n$ if, for every x_o, there exist a convex neighborhood V of x_o, convex functions g_V and h_V such that :

$$f(x) = g_V(x) - h_V(x) \text{ for all } x \in V. \tag{2.1}$$

THEOREM 2.2. [HARTMAN, 1959]. *Every locally d.c. function on $\mathbb{R}^n$ is globally d.c. on $\mathbb{R}^n$.*

HARTMAN actually stated his theorem in a slightly more general form, by extending locally d.c. functions on a closed (or open) convex set. This type of result, although well known in the convex case, is rather surprising since it means that information on f around each point is enough to decide that f is globally d.c. on $\mathbb{R}^n$. HARTMAN's proof needs to be "freed from dust" somewhat ; the extension procedures HARTMAN uses can be adapted by relying on extension techniques from modern convex analysis (infimal convolution of a function with $k\|\cdot\|$ for example). As a mere consequence of HARTMAN's theorem, we note that *every function f which is C^2 on $\mathbb{R}^n$ is d.c. on $\mathbb{R}^n$.* It is easy to see that such a function is locally d.c.. Indeed, due to the continuity of $\nabla^2 f$, one can find a decomposition of f on $\overline{B}(\bar{x},r)$ as :

$$f(x) = (f(x) + \rho\|x\|^2) - \rho\|x\|^2,$$

where ρ is chosen such that $f + \rho\|\cdot\|^2$ be convex on $\overline{B}(\bar{x},r)$. Hence, for every C^2 function f, there are convex functions g and h such that :

$$f(x) = g(x) - h(x) \text{ for all } x \in \mathbb{R}^n. \tag{2.3}$$

Moreover, it has been proved by BOUGEARD ([7]) that the function h in the decomposition above could be chosen to be C^∞. Hence *any* C^2 *function on* $\mathbb{R}^n$ *can be written as a difference of two convex functions, one of them being* C^2 *and the other one* C^∞. POMMELLET ([37]) also offered an alternative proof that a C^2 function could be written as a difference of two C^2 convex functions. The basic idea to prove such type of results is to build up global decompositions starting from local decompositions and using extension (or regularization) procedures of some kind. However, all the proofs we know are "constructive" in the sense that they indeed yield g and h satisfying (2.3) but could hardly be carried over computational aspects.

Quite a little is missing for a C^1 function to be d.c.. Actually, a $C^{1,1}$ function, that is a function f whose gradient is locally Lipschitz, is d.c.. $C^{1,1}$ functions constitute a subclass of the so-called *lower-*C^2 *functions*. The class of lower-C^2 functions has been studied in the literature by several authors, using different names and apparently unaware that the functions they were talking about belong to the same class (cf. [26], [29], [42], [53], [54], etc...). Among the various characterizations of lower-C^2 functions, we retain the following one : f is lower-C^2 on $\mathbb{R}^n$ if for every x_o, there exist a convex neighborhood V of x_o, a convex function g_V, a quadratic convex function h_V such that :

$$f(x) = g_V(x) - h_V(x) \text{ for all } x \in V. \qquad (2.4)$$

The class of lower-C^2 functions on $\mathbb{R}^n$ is denoted by $LC^2(\mathbb{R}^n)$. This class remains unchanged if we impose that h_V in decomposition (2.4) be $C^{1,1}$ (or C^2, or C^∞) and convex. We clearly have :

$$\mathrm{Conv}(\mathbb{R}^n) \subset LC^2(\mathbb{R}^n) \subset DC(\mathbb{R}^n) \quad ; \qquad (2.5)$$

$$C^{1,1}(\mathbb{R}^n) \subset LC^2(\mathbb{R}^n). \qquad (2.6)$$

$LC^2(\mathbb{R}^n)$ is an example of convex cone of functions getting in the framework announced in (0.2). The vector space generated by $LC^2(\mathbb{R}^n)$ *is* $DC(\mathbb{R}^n)$ while the largest vector space $\mathcal{L}$ contained in $LC^2(\mathbb{R}^n)$ (that is : $\mathcal{L} = LC^2(\mathbb{R}^n) \cap - LC^2(\mathbb{R}^n)$) contains $C^{1,1}(\mathbb{R}^n)$. Actually the f belonging to $\mathcal{L}$ are characterized as follows : f is differentiable and for all $x_o \in \mathbb{R}^n$ there is a neighborhood V of x_o and a positive k such that :

$$|\langle \nabla f(x) - \nabla f(x'), x-x' \rangle| \leq k \|x-x'\|^2 \text{ whenever } x,x' \in V. \qquad (2.7)$$

Hence $\mathcal{L}$ *contains* $C^{1,1}(\mathbb{R}^n)$ and is *included* in $C^1(\mathbb{R}^n)$. f is said to be *globally lower-*C^2 on $\mathbb{R}^n$ if the neighborhood V in the definition (2.4) is imposed to be the whole of $\mathbb{R}^n$. Contrary to what happens for the d.c. character, a lower-C^2 function on $\mathbb{R}^n$ is not necessarily globally lower-C^2 on $\mathbb{R}^n$.

To see there is a strong gap between $LC^2(\mathbb{R}^n)$ and $DC(\mathbb{R}^n)$, we consider again the function d_S^2 (cf. example 0.2).

PROPOSITION 2.3. *Let S be a nonempty closed set in* $\mathbb{R}^n$. *Then the function* d_S^2 *is lower-*C^2 *if and only if S in convex.*

Proof. d_S^2 is convex, hence lower-C^2, whenever S is convex. Due to an alternative characterization of lower-C^2 functions by ROCKAFELLAR ([42]), one easily shows that $-d_S^2$ is lower-C^2 on $\mathbb{R}^n$. So, assuming that d_S^2 itself is lower-C^2, we are in the presence of a differentiable function d_S^2. We know from ASPLUND (1973) that the function $h : x \to \frac{1}{2}\|x\|^2 - \frac{1}{2}d_S^2(x)$ is convex and that the subdifferential $\partial h(x_0)$ contains $P_S(x_0) = \{u \in S ; \|x_0 - u\| = d_S(x_0)\}$. Thus, in our case, $P_S(x)$ is a singleton for all x. That means that S is a so-called CHEBYSHEV set, and, in $\mathbb{R}^n$, the only CHEBYSHEV sets are convex sets. □

II.2. Lower-C^2 functions can be characterized via their generalized gradients. Following ROCKAFELLAR's terminology ([42]), the generalized gradient ∂f of a locally Lipschitz function $f : \mathbb{R}^n \to \mathbb{R}$ is said to be *strictly hypomonotone* at x_0 if there exists a neighborhood V of x_0 and a positive k such that :

$$\langle x-x', y-y' \rangle \geq -k \|x-x'\|^2 \text{ for all } x,x' \text{ in } V$$
$$\text{and } y \in \partial f(x),\ y' \in \partial f(x'). \tag{2.8}$$

ROCKAFELLAR's characterization of a lower-C^2 function via its generalized gradient comes as follows : *a locally Lipschitz* $f : \mathbb{R}^n \to \mathbb{R}$ *is lower-*C^2 *if and only if, for each* $x_0 \in \mathbb{R}^n$, ∂f *is strictly hypomonotone at* x_0. One may wonder whether there is a similar characteristic property for d.c. functions. Simple examples show that such a characterization is by no means easy to derive. The function f in Example 1.4 has a generalized gradient "varying very nicely" since $\partial f(x) = [0,1]$ for all x ; in imprecise terms, ∂f is of "bounded variation" on $\mathbb{R}$ (the HAUSFORFF distance between $\partial f(x)$ and $\partial f(x')$ is null for all x,x'). The one-dimensional case suggests us to look at a notion of "bounded variation" for functions of several variables. There are, in the literature, at least half a dozen different definitions for "$F : \Omega \subset \mathbb{R}^n \to \mathbb{R}^m$ is of bounded variation on Ω". Grafting them to our situation has not yielded very significant results ; in particular, the following desired result could not be obtained : a differentiable function $f : \mathbb{R}^n \to \mathbb{R}$ is d.c. if and only if its gradient mapping ∇f is of bounded variation on a neighborhood of each point. This is not, after all, so surprising. Each definition of "bounded variation" has been introduced for particular purposes (e.g., the concept of bounded variation or bounded deformation in the theory of plasticity), while the notion we are looking for has to be strongly connected to that of monotonicity such as used in the context of Convex Analysis. The next definition is an attempt by ELLAIA ([18, chapter II]) to cope with that problem.

DEFINITION 2.4. *A multifunction $\Gamma : \mathbb{R}^n \rightrightarrows \mathbb{R}^n$ is of bounded variation at $x_0 \in \mathbb{R}^n$ if there exists a maximal cyclically monotone multifunction $M : \mathbb{R}^n \rightrightarrows \mathbb{R}^n$ containing x_0 in the interior of its domain, such that :*

$$\limsup_{\substack{x \to x_0,\ x' \to x_0 \\ y \in \Gamma(x),\ y' \in \Gamma(x') \\ z \in M(x),\ z' \in M(x') \\ \langle x-x', z-z'\rangle \neq 0}} \frac{|\langle x-x', y-y'\rangle|}{\langle x-x', z-z'\rangle} < +\infty. \tag{2.9}$$

In other words, for some positive k and a neighborhood V of x_0,

$$|\langle x-x', \Gamma(x) - \Gamma(x')\rangle| \leq k \langle x-x', M(x) - M(x')\rangle \quad \text{for all } x, x' \text{ in } V. \tag{2.10}$$

This is actually a requirement on the "angles" $\langle x-x', \Gamma(x) - \Gamma(x')\rangle$ and $\langle x-x', M(x) - M(x')\rangle$. This concept is easier to grasp when Γ is merely a mapping ; of course, when n = 1, it amounts to the usual notion of bounded variation in a neighborhood of x_0. If Γ_1 and Γ_2 are of bounded variation at x_0, so is $\lambda\Gamma_1 + \mu\Gamma_2$ for any $\lambda, \mu \in \mathbb{R}$.

Due to what has been assumed on M, M is the subdifferential of a convex function finite in a neighborhood of x_0. So, the next result was foreseeable.

<u>THEOREM 2.5</u> ([18]). *A locally Lipschitz function $f : \mathbb{R}^n \to \mathbb{R}$ is d.c. if and only if ∂f is of bounded variation at all $x_0 \in \mathbb{R}^n$.*

The definition of bounded variation we have considered for ∂f is not quite satisfactory ; in particular it is heavy to use. More work should be done to get a clear-cut characterization of d.c. functions via their first-order (generalized) derivatives.

<u>II.3</u>. Since C^2 functions on $\mathbb{R}^n$ are always d.c., little room is left for characterizing (non C^2) d.c. functions f when having some (generalized) second derivative of f at our disposal. An approach which can be carried out to a certain extent is that of considering *second derivatives of f in the distributional sense*. When f is defined on the real line ($n = 1$), it is known that f is d.c. if and only if the second derivative of f is a RADON measure. When $n > 1$, some *necessary* conditions can be derived from DUDLEY's results on convex functions ([14]) ; for example, if $f \in DC(\mathbb{R}^n)$, the second derivative of the distribution associated with f is a (n,n) matrix-valued RADON measure. For the converse, decomposing a (n,n) matrix-valued RADON measure into the difference of two nonnegative (n,n) matrix-valued RADON measures does not help very much since one does not know whether the resulting distributions are second derivatives. The same type of difficulties will arise in decomposing d.c. functions (see Section III).

When pointwise derivatives of f are available, again some necessary conditions can be derived in terms of the total Gaussian curvature of f ([54]).

III. DECOMPOSING D.C. FUNCTIONS

Given $f \in DC(\mathbb{R}^n)$, it is evident there are infinitely many manners of expressing it as a difference of convex functions. Is there somehow a way of reducing the choice of such convex functions ? How to define that a decomposition is better than another one ? Finally, from the practical viewpoint, how to find a decomposition of a function f which has been constructed from other functions whose decompositions are better known ? The present section addresses to these questions.

Firstly, let us remark that if a decomposition $f = g-h$ is available, one can always choose g and h as being *strictly* (*or uniformly*) *convex* ; it suffices to write :

$$f = (g+\varphi) - (h+\varphi) = \bar{g} - \bar{h}, \tag{3.1}$$

with φ strictly (resp. uniformly) convex. The simplest choice consists in taking $\varphi(x) = k\,\|x\|^2$ with $k > 0$. Adding functions like $k\,\|\cdot\|^2$ may be deliberate since it adds more structure to the functions $\bar{g}$ and $\bar{h}$ involved in the decomposition of f ; in particular - and this is of importance in the context of duality (see next section) - it ensures that both $(\bar{g})^*$ and $(\bar{h})^*$ are everywhere finite and differentiable.

We say that a decomposition $f = g - h$ is *normalized* if $\inf_{\mathbb{R}^n} h(x) = 0$.

PROPOSITION 3.1. *Every d.c. function enjoys normalized decompositions.*

Proof. Given $h \in \mathrm{Conv}(\mathbb{R}^n)$, we show there is an affine function θ such that $\inf_{\mathbb{R}^n} (h+\theta) = 0$.

Let $x_0^* \in \mathrm{dom}\, h^*$. We pose $\theta(x) = -\langle x_0^*, x\rangle + h^*(x_0^*)$. Thus

$$(h+\theta)^*\ (x^*) = -\,h^*(x_0^*) + h^*(x^* + x_0^*) \text{ for all } x^*.$$

Whence $(h+\theta)^*(0) = 0$. □

III.1. A normalized decomposition $f = g_{min} - h_{min}$ is called *minimal* if $g_{min} \le g$ (and, consequently, $h_{min} \le h$) whatever $f = g-h$ be a normalized decomposition. This definition is very stringent since it requires g and h to be comparable with g_{min} and h_{min} whatever g and h appear in a normalized decomposition of f. A weaker condition would consist in requiring $g_{min} \le g$ whenever g_{min} and g are comparable. Nevertheless, apart from some particular cases (like d.c. functions on $\mathbb{R}$ or polyhedral functions on $\mathbb{R}^n$), there is no "minimal" decomposition as long as the pointwise ordering of functions ($\varphi \le \psi$ if $\varphi(x) \le \psi(x)$ for all x) is considered. There are counterexamples even for d.c. functions of two variables.

Example 3.1. (from [21])

Let $f \in DC(\mathbb{R}^2)$ be defined as $f(\xi_1, \xi_2) = 2\,\xi_1\,\xi_2$.

The following are normalized decompositions of f :

$$f(\xi_1, \xi_2) = g_\varepsilon(\xi_1, \xi_2) - h_\varepsilon(\xi_1, \xi_2),$$

where

$$g_\varepsilon(\xi_1, \xi_2) = (\varepsilon\xi_1 + \frac{\xi_2}{\varepsilon})^2,$$
$$h_\varepsilon(\xi_1, \xi_2) = \varepsilon^2 \xi_1^2 + \frac{\xi_2^2}{\varepsilon^2}, \tag{3.2}$$

and $\varepsilon > 0$. Observe the antagonistic role played by ε^2 and $1/\varepsilon^2$.

If a minimal (normalized) decomposition $f = g_{min} - h_{min}$ should exist, we would have :

$$0 \leq h_{min}(\xi_1, \xi_2) \leq \varepsilon^2 \xi_1^2 + \frac{\xi_2^2}{\varepsilon^2} \text{ for all } (\xi_1, \xi_2) \in \mathbb{R}^2$$

and all $\varepsilon > 0$.

Since h_{min} is convex, the above implies that $h_{min} \equiv 0$. Thus, $f = g_{min}$, which is impossible because f is not convex on $\mathbb{R}^2$.

Let us go back to the example of quadratic forms on $\mathbb{R}^n$ (cf. example 0.1)

$$Q(x) = \frac{1}{2} < Ax, x >.$$

$g : x \to g(x) = \frac{1}{2}\{<Ax,x> + k \|x\|^2\}$ is convex for k large enough, so that Q can be decomposed as :

$$Q(x) = \frac{1}{2}\{<Ax,x> + k\|x\|^2\} - \frac{1}{2} k \|x\|^2. \tag{3.3}$$

Another possible decomposition of Q is via the diagonalization D of A. Let D^+ (resp. D^-) denote the diagonal matrix whose elements are positive parts (resp. negative parts) of D. The decomposition $D = D^+ - D^-$ yields a decomposition of Q,

$$Q(x) = \frac{1}{2} <A^+ x,x> - \frac{1}{2} < A^- x,x>, \tag{3.4}$$

where A^+ and A^- are positive semi-definite (and singular if Q is neither convex nor concave). Is decomposition (3.4) "better" than decomposition (3.3) ? Is decomposition (3.4) optimal in some sense ? This example just as the previous ones clearly show that *minimal decompositions have to be searched for in restricted classes of decompositions*. So, a more sensible question for the decomposition of quadratic forms would be : is decomposition (3.4) minimal in the class of quadratic decompositions of Q ? When trying to decompose a C^2 function as a difference of convex functions, one might be tempted to decompose $\nabla^2 f(x)$ as it is done in the scheme leading to (2.4) for example ;

$$\langle \nabla^2 f(x)d,d \rangle = \langle A^+(x)d,d \rangle - \langle A^-(x)d,d \rangle. \tag{3.5}$$

The drawback, foreseen in the previous section, is that the $A^+(x)$ and $A^-(x)$ obtained in such a way are not necessarily second derivatives of functions! The case of $f(\xi_1, \xi_2) = \xi_1/\xi_2$ on $(\mathbb{R}_+^*)^2$ is a typical counterexample.

In sum, given a function $f \in DC(\mathbb{R}^n)$, there is no general rule for finding in an automatic way a decomposition of f as a difference of convex functions.

III.2. As mentioned earlier, $DC(\mathbb{R}^n)$ is closed under all the operations usually considered in Optimization. So, given d.c. functions f_i whose decompositions $g_i - h_i$ are known, how to find a decomposition $g - h$ of a d.c. function f constructed from the f_i ? The problem is not as difficult as it might appear at the first glance ; that depends, of course, of the operation carried out on the f_i.

If f is a linear combination of the f_i, no trouble arises since a decomposition of f comes out immediately. If now f is the maximum (or the minimum) of the f_i, one gets a decomposition of f as follows :

$$\max_{i=1,\dots,k} f_i = \max_{i=1,\dots,k} \{g_i + \sum_{\substack{j=1\\ j\neq i}}^{k} h_j\} - \sum_{i=1}^{k} h_i \;; \tag{3.6}$$

$$\min_{i=1,\dots,k} f_i = \sum_{i=1}^{k} g_i - \max_{i=1,\dots,k} \{h_i + \sum_{\substack{j=1\\ j\neq i}}^{k} g_j\} . \tag{3.7}$$

As for example, if a decomposition $g - h$ of f is available, we have :

$$\begin{aligned} |f| &= 2 \max (g,h) - (g+h) \;; \\ f^+ &= \max (g,h) - h \;; \\ f^- &= \max (g,h) - g. \end{aligned} \tag{3.8}$$

As for the product of two d.c. functions f_1 and f_2, we firstly decompose f_i as $f_i^+ - f_i^-$ so that the question reduces to finding a decomposition of the product of two *positive* d.c. functions. Let therefore $f_1 = g_1 - h_1$

and $f_2 = g_2 - h_2$ be normalized decompositions of positive d.c. functions f_1 and f_2. The g_i and h_i are thus positive convex functions. A decomposition of $f_1 \cdot f_2$ is then at hand :

$$f_1 \cdot f_2 = \frac{1}{2}[(h_1+h_2)^2+(g_1+g_2)^2] - \frac{1}{2}[(h_1+g_1)^2+(h_2+g_2)^2]. \qquad (3.9)$$

By reiterating or combining such decomposition rules, one often is able to find a decomposition of a function which has been constructed from other functions whose decompositions are better known. See, for instance, the decomposition of the d.c. function f arising in Example 1.1.

IV. OPTIMALITY CONDITIONS, DUALITY FOR D.C. FUNCTIONS

IV.1. *Optimality conditions*

D.c. functions are locally Lipschitz and possess directional derivatives ; optimality conditions for d.c. mathematical programs can therefore be deduced from those derived in the context of mathematical programs with directionally differentiable data ([36], [39]), or tangentially d.c. data ([12], [45], [46]), or merely locally Lipschitz data ([11]). However, knowing that the data can be written as differences of convex functions adds to the structure of the problem and may be used for optimality conditions. To begin with, consider the unconstrained case :

$$(\mathcal{P}) \text{ Minimize } f(x) = g(x) - h(x) \text{ over } \mathbb{R}^n.$$

A necessary condition for x_0 to be a local minimum of f is that :

$$0 \in \partial f(x_0) \subset \partial g(x_0) - \partial h(x_0) . \qquad (4.1)$$

In other words, the subdifferentials $\partial g(x_0)$ and $\partial h(x_0)$ must overlap :

$$\partial g(x_0) \cap \partial h(x_0) \neq \phi . \qquad (4.2)$$

The same condition holds true when x_0 is a local maximum of f. The major drawback of (4.2) is that it *depends on the decomposition of f* as a difference of convex functions g and h (besides the fact that it is not always informative). In particular - and this is also true for the definition of critical points of f (see paragraph IV.3) - a given x_0 may satisfy condition (4.2) for an ad hoc decomposition of f. A further approach consists in expressing necessary conditions for optimality via the *lower-subdifferen-*

tial of f in the sense of PENOT ([36]). If x_o is a local minimum of f, then

$$f'(x_o;d) \geq o \text{ for all } d \in \mathbb{R}^n, \tag{4.3}$$

so that :

$$o \in \underline{\partial} f(x_o), \tag{4.4}$$

where $\underline{\partial} f(x)$ stands for :

$$\{x^* \in \mathbb{R}^n \mid \langle x^*,d \rangle \leq f'(x;d)\}. \tag{4.5}$$

There is an alternate way of looking at the optimality condition (4.4), that is by way of the $\star$-difference of $\partial g(x_o)$ and $\partial h(x_o)$. Given two nonempty compact convex sets A and B, the $\star$-difference of A and B is the set

$$A \underline{\star} B = \{x \in \mathbb{R}^n \mid x + B \subset A\}.$$

This operation was introduced in the context of linear differential games by PONTRYAGIN and further exploited by PSHENICHNYI ([38]) for convex optimization. If the algebraic difference A - B of A and B is known to be "too large", the $\star$-difference turns out to be often "too small" ! $A \underline{\star} B$ is actually a compact convex set whose support function is the biconjugate of the difference of the support functions of A and B ;

$$\psi^*_{A \underline{\star} B} = (\psi^*_A - \psi^*_B)^{**}. \tag{4.6}$$

We infer from (4.3) that the biconjugate function of $d \to f'(x_o;d) = g'(x_o;d) - h'(x_o;d)$ is positive. Whence a necessary condition for x_o to be a local minimum of $f(x) = g(x) - h(x)$ is that :

$$o \in \partial g(x_o) \underline{\star} \partial h(x_o). \tag{4.7}$$

Observe that $\partial g(x_o) \underline{\star} \partial h(x_o)$ does not depend on g and h but on the difference g - h. Finally, note that :

$$\underline{\partial} f(x_o) = \partial g(x_o) \underline{\star} \partial h(x_o),$$

so that (4.7) is a reformulation of the optimality condition (4.4). To summarize, a necessary condition for x_o to be a local minimum of f is that :

$$o \in \partial g(x_o) \underline{\star} \partial h(x_o) \subset \underline{\partial} f(x_o) \subset \partial g(x_o) - \partial h(x_o). \tag{4.8}$$

A *sufficient* condition for x_o to be a (strict) local minimum of f can also be expressed in terms of $\partial g(x_o) \underline{*} \partial h(x_o)$ by just exploiting the sufficient condition for minimality :

$$f'(x_o;d) > o \text{ for all non-null } d. \tag{4.9}$$

The following is easy to prove : *if* o *lies in the interior of* $\partial g(x_o) \underline{*} \partial h(x_o)$, *then* x_o *is a strict local minimum of* f.

Consider now the constrained minimization problem :

$$(\mathcal{P}) \text{ Minimize } f(x) = g(x) - h(x) \text{ over } S.$$

The intended work, in its generality, would consist in deriving optimality conditions when the constraint set is represented via inequalities and equalities involving d.c. functions. That has been done to a certain extent in [12], [18, p. 95-100], [46]. We show here how the above-displayed conditions extend to the constrained case, when the constraints are formulated as : $x \in S$. For $x_o \in S$, we recall that the *contingent cone* to S at x_o (or BOULIGAND's tangent cone to S at x_o) is the closed cone with apex 0, denoted by $T(S;x_o)$ and defined as :

$$T(S;x_o) = \{d \mid \exists\, (d_n) \to d,\ \exists (\alpha_n) \to o^+ \text{ such that } x_o + \alpha_n d_n \in S \text{ for all } n\}.$$

<u>THEOREM 4.1</u>. *Let* $x_o \in S$ *be a local minimum of* f *on* S, *let* T *be a closed convex cone included in* $T(S;x_o)$. *Then :*

$$o \in [\partial g(x_o) + T^\circ] \underline{*} \partial h(x_o) . \tag{4.10}$$

Let now N *be a closed convex cone included in* $T(S;x_o)^\circ$. *If*

$$o \in \mathit{int}\{[\partial g(x_o) + N] \underline{*} \partial h(x_o)\}, \tag{4.11}$$

then x_o *is a strict local minimum of* f *on* S.

Proof. If x_o is a local minimum of f on S, then $f'(x_o;d) \geq 0$ for all $d \in T(S;x_o)$. This implies that :

$$f'(x_o;d) + \psi_T(d) \geq 0 \text{ for all } d,$$

where $\psi_T = \psi^*_{T^\circ}$ is the indicator function of T (or the support function of T°). Consequently

$$g'(x_o;d) + \psi^*_{T^\circ}(d) \geq h'(x_o;d) \text{ for all } d.$$

In equivalent terms,

$$\psi^*_{\{\partial g(x_o) + T^\circ\}}(d) \geq \psi^*_{\partial h(x_o)}(d) \text{ for all d.}$$

When the announced condition (4.10), by just extending the definition of $A \underline{*} B$ to the case where A is not bounded.

If $N \subset T(S;x_o)^\circ$ and if condition (4.11) is satisfied, we have :

$$g'(x_o;d) + \psi^*_{T(S;x_o)^\circ}(d) > h'(x_o;d) \text{ for all non-null d,}$$

so that

$$f'(x_o;d) > 0 \text{ for all non-null d in } T(S;x_o). \tag{4.12}$$

Suppose there is a sequence $(x_n) \subset S$ converging to x_o, $x_n \neq x_o$, such that $f(x_n) \leq f(x_o)$. Subsequencing if necessary, we may suppose that

$$\frac{x_n - x_o}{\|x_n - x_o\|}$$

does converge to a limit $\overline{d}$, $\overline{d} \neq 0$. $\overline{d}$ belongs to $T(S;x_o)$ since

$$x_n = x_o + \|x_n - x_o\| \overline{d} \text{ converges to } x_o \text{ in S.}$$

The inequality $f(x_n) \leq f(x_o)$ makes that $f'(x_o;\overline{d}) \leq 0$, which is in contradiction with (4.12). □

IV.2. *Duality*

Given $f = g-h \in DC(\mathbb{R}^n)$, the conjugate f^* of f can be expressed in terms of the conjugate of the (convex) functions g and h. Duality schemes involving d.c. functions actually take root in this basic formula we state now.

THEOREM 4.2. *The conjugate of $f = g-h$ is given as :*

$$\forall\, x^* \in \mathbb{R}^n \quad f^*(x^*) = \sup_{y^* \in \operatorname{dom} h^*} \{g^*(x^* + y^*) - h^*(y^*)\}, \tag{4.13}$$

where $\operatorname{dom} h^* = \{x^* \mid h^*(x^*) < +\infty\}$.

Observe the symmetry with the formula giving the conjugate of (g + h) :

$$(g+h)^*(x^*) = \inf_{y^* \in \operatorname{dom} h^*} \{g^*(x^*-y^*) + h^*(y^*)\} . \tag{4.14}$$

The two formulae (4.13) and (4.14) overlap only when h is an affine function.

Since $-f^*(o) = \inf_{x \in \mathbb{R}^n} f(x)$, we deduce from (4.13) :

$$\inf_{x \in \mathbb{R}^n} \{g(x) - h(x)\} = \inf_{x^* \in \operatorname{dom} h^*} \{h^*(x^*) - g^*(x^*)\}. \tag{4.15}$$

By just using that $\sup(\cdot) = -\inf(-\cdot)$, we also derive :

$$\sup_{x \in \mathbb{R}^n} \{g(x) - h(x)\} = \sup_{x^* \in \operatorname{dom} g^*} \{h^*(x^*) - g^*(x^*)\} . \tag{4.16}$$

Theorem 4.2. is due to PSHENICHNYI ([38]). As an application, he proved that the conjugate of the difference of two support functions, $\psi_A^* - \psi_B^*$, is precisely the indicator function of $A \underline{*} B$. The result (4.13) remains true when g is an arbitrary function taking possibly the value $+ \infty$ (cf. [19] for a short proof) ; this is of importance in order to include the constraint in the objective function when building up associated dual problems.

As for an example, consider the problem of maximizing a convex function h over a convex set S :

$$\alpha = \sup_{x \in S} h(x). \tag{4.17}$$

We clearly have that :

$$- \alpha = \inf_{x \in \mathbb{R}^n} \{\psi_S(x) - h(x)\}.$$

It comes from (4.15) that

$$- \alpha = \inf_{x^* \in \operatorname{dom} h^*} \{h^*(x^*) - \psi_S^*(x^*)\}. \tag{4.18}$$

To go further in this example, suppose $h(x) = \frac{1}{2} < Ax, x >$, where A is a symmetric positive definite (n,n) matrix and S the euclidean unit ball. Then 2α is the largest eigenvalue λ_M of A and we deduce from (4.18) the dual formulation announced in Example 0.3, that is :

$$- \alpha = - \lambda_M/2 = \inf_{x^* \in \mathbb{R}^n} \{ \|x^*\| - \frac{1}{2} < A^{-1} x^*, x^*> \} . \qquad (4.19)$$

Another variational formulation for λ_M is :

$$\alpha = \lambda_M/2 = \sup \{ \frac{1}{2} \|x\|^2 ; <A^{-1} x, x> \le 1 \}.$$

$\frac{1}{2}\|\cdot\|^2$ equals its conjugate function while the support function of the elliptic set $\{x | <A^{-1}x,x> \le 1\}$ is $x^* \to \sqrt{<Ax^*,x^*>}$. Consequently, we infer from (4.18) the alternate formulation (0.6) given in example 0.3..

In successive papers ([49], [50], [51]), TOLAND proved and exploited thoroughly the equality (4.15). The mapping which assigns $h^* - g^*$ to $g - h$ is sometimes referred to as *TOLAND's involution*. The duality schemes TOLAND proposed in the context of Calculus of Variations have been further developed by AUCHMUTY ([6]). Relationships with LEGENDRE's transformation are laid out in EKELAND's papers ([15], [16]). A formula like (4.13) in its most general and abstract setting has recently been derived by the author ([24]).

A further interesting consequence of relation (4.15) concerns with the *regularization* of f. Given the convex function g or h, a natural way of regularizing (and, possibly, "smoothing") g is by performing the infimal convolution of g with some "kernel" θ :

$$g \nabla \theta : x \to (g \nabla \theta)(x) = \inf_{y \in \mathbb{R}^n} \{g(y) + \theta(x-y)\} . \qquad (4.20)$$

Two important examples are :

$$g_r = g \nabla \frac{1}{2r} \|\cdot\|^2, \; r > 0 \quad ;$$
$$\overline{g}_r = g \nabla \frac{1}{r} \|\cdot\| , \; r > 0 \qquad (4.21)$$

Each of these regularization procedures has its own advantages. The first one, widely used in Nonlinear Analysis (the so-called MOREAU-YOSIDA regularization scheme) or Convex Optimization (algorithms based upon the so-called proximal mapping), gives rise to a $C^{1,1}$ function g_r which coincides with g only at minimumpoints. The second one is also of interest since it gives rise to a Lipschitz function (for r small enough) which coincides with g at those points x where $\partial g(x)$ and the ball $\overline{B}(o,1/r)$ overlap. Since $(g \nabla \theta)^* = g^* + \theta^*$ in the examples we have considered, we deduce from (4.15) that :

$$\inf_{x \in \mathbb{R}^n} f(x) = \inf_{x \in \mathbb{R}^n} \{g(x) - h(x)\} = \inf_{x \in \mathbb{R}^n} \{g_r(x) - h_r(x)\}. \tag{4.22}$$

$$= \inf_{x \in \mathbb{R}^n} \{\overline{g}_r(x) - \overline{h}_r(x)\}. \tag{4.23}$$

The relationship (4.22) was noticed by GABAY ([20]) who used it for algorithmic purposes.

IV.3. *Comparing critical points of $g-h$ and h^*-g^**

The first question which arises concerning critical points of $f = g-h$ is that of the *definition* itself. According to TOLAND ([49], [51]), x_0 is a critical point of f if $\partial g(x_0) \cap \partial h(x_0) \neq \phi$ (cf. (4.1) and (4.2)). This definition is not quite satisfactory since, as stated earlier, it depends on the decomposition of f as a difference of convex functions g and h. Another definition, drawing inspiration from (4.7), comes as follows ([18, ch. II]) :

DEFINITION 4.3. *x_0 is said to be a lower critical point of $f = g-h$ (resp. an upper critical point of f) if :*

$$o \in \partial g(x_0) \stackrel{*}{-} \partial h(x_0) \quad (\textit{resp. } o \in \partial h(x_0) \stackrel{*}{-} \partial g(x_0)).$$

This definition is stringent ; it however does not depend on g and h in the decomposition $g-h$ of f. Apart from the case where f is differentiable, either $\partial g(x_0) \stackrel{*}{-} \partial h(x_0)$ or $\partial h(x_0) \stackrel{*}{-} \partial g(x_0)$ is empty.

Quite interesting relationships between critical points of $g-h$ and $h^* - g^*$ have been proved by TOLAND ([49], [51]). We state here an example of such comparison results.

THEOREM 4.4. *If x_0 is a critical point of $g-h$, then every $x_0^* \in \partial g(x_0) \cap \partial h(x_0)$ is a critical point of $h^* - g^*$ and $g(x_0) - h(x_0) = h^*(x_0^*) - g^*(x_0^*)$.*

If x_0 is a global minimum of $g-h$, then every $x_0^* \in \partial h(x_0)$ is a global minimum of $h^* - g^*$.

$h^* - g^*$ is unambiguously defined by adopting the addition rule $\infty - \infty = \infty$.

Counterparts with lower (or upper) critical points can be derived to a certain extent. Usable results however do not go beyond TOLAND's ones.

Counterexample (PENOT).

Let $f \in DC(\mathbb{R}^n)$ be defined as the difference of the support functions of A and B :

$$f = \psi_A^* - \psi_B^*$$
$$= g - h ,$$

with $A = [-1,1] \times \{o\}$ and B the closed unit ball in $\mathbb{R}^n$. $x_o = (\alpha, o), \alpha > o$, is a lower critical point of f since $\partial g(x_o) = \partial h(x_o) = \{(1,o)\}$. But for $x_o^* = (1,o)$, $\partial g^*(x_o^*) = \mathbb{R}_+ \times \mathbb{R}$ and $\partial h^*(x_o^*) = \mathbb{R}_+ \times \{0\}$. So, x_o^* is a critical point of $h^* - g^*$ in TOLAND's sense but is no more a lower critical point.

Further classifications of critical points using "second-order information" on g and h rely on MORSE theory for d.c. functions. For that, peruse [7] and [37].

V. PREVIEW ON MINIMIZATION PROCEDURES FOR D.C. FUNCTIONS

Minimizing (globally) a d.c. function is strongly related to some other difficult optimization problems like that of finding the global maximum of a convex function over a convex set. We have seen in IV.2. how the problem of *maximizing a convex function over a convex set* could be transformed into that of *minimizing a d.c. function over the whole space*. Conversely, consider the following d.c. minimization problem :

$$(\mathcal{P}) \text{ Minimize } f(x) = g(x) - h(x) \text{ over } S,$$

where $g, h \in \text{Conv}(\mathbb{R}^n)$ and S is convex.

Define $\overline{g}, \overline{h} : \mathbb{R}^{n+1} \to \mathbb{R}$ by $\overline{g}(x,\xi) = g(x) - \xi$ and $\overline{h}(x,\xi) = h(x) - \xi$. Consider now :

$$(\mathcal{Q}) \text{ Maximize } \overline{h}(x,\xi) \text{ subject to } \overline{g}(x,\xi) \leq 0 \text{ and } (x,\xi) \in S \times \mathbb{R}.$$

($\mathcal{Q}$) is a maximization problem of a convex function over a convex set. ($\mathcal{P}$) and ($\mathcal{Q}$) are related in the following way : *$(x_0, \xi_0 = g(x_0))$ is a solution of ($\mathcal{Q}$) if and only if x_0 is a solution of ($\mathcal{P}$).*

A further example giving rise to a d.c. minimization problem comes from *fractional programming*. Consider the so-called "convex-convex fractional program" :

$$(\mathcal{F})\ \text{Maximize } q(x) = \frac{g(x)}{h(x)} \text{ over } S,$$

where both g and h are convex (h strictly positive) and S is a compact convex set. An old approach in fractional programming (DINKELBACH, 1967) consists in transforming ($\mathcal{F}$) into a parametric d.c. program :

$$(\mathcal{F}_r)\ \text{Minimize } \{rh(x) - g(x)\} \text{ over } S,\ r \in \mathbb{R}.$$

Denote by $\bar{r}$ the unique zero of the strictly increasing function

$$r \to \bar{q}(r) = \text{Min } \{rh(x) - g(x) \mid x \in S\}.$$

Then, *a solution of ($\mathcal{F}_r$) with $r = \bar{r}$ is also a solution of ($\mathcal{F}$).*

Convexity is present twice (in g and h) for any decomposition of a d.c. function $f = g - h$. One may think of using alternately $\partial g(\cdot)$ and $\partial h(\cdot)$ to generate a sequence (x_n) which would converge to some local (global ?) minimum of f. Most of the attempts using this "first-order information" $\partial g(x_n)$, $\partial h(x_n)$ at the current point x_n lead to sequences which converge to stationary or critical points (of some kind) of f. Quite a little has been done to devise algorithms for a *global* minimization of d.c. functions ([35], [52]). This is a promising and important area of research, although one should not expect miracles...

CONCLUSION

Concerning the analysis and the optimization of d.c. functions, the main contributions are either old (around the thirties for Analysis) or quite recent (1979 - now for what concerns Optimization). The present survey reflects the thinking of the author while guiding ELLAIA's thesis ([18]).

It seems to us there is a growing interest in studying d.c. minimization problems, the main incentive coming from modelling in Applied

Mathematics. That induced H. TUY and the author to undertake the edition of a collection of research works in this field. That should be viewed as a natural extension of the present survey.

REFERENCES

[1] A.D. ALEXANDROFF, *Almost everywhere existence of the second differential of a convex function and some properties of convex surfaces connected with it*, Ucenye Zapiski Leningr. Gos. Univ. Ser. Mat. 37 n° 6, (1939) 3-35 (in Russian)

[2] A.D. ALEXANDROFF, *On surfaces represented as the difference of convex functions*, Izv. Akad. Nauk Kaz. SSR 60, Ser. Mat. Mekh. 3, (1949) 3-20 (in Russian).

[3] A.D. ALEXANDROFF, *Surfaces represented by the differences of convex functions*, Dokl. Akad. Nauk SSSR 72, (1959) 613-616 (in Russian).

[4] G. ARSOVE, *Functions representable as the difference of subharmonic functions*, Transactions Amer. Math. Soc. 75, (1953) 327-365.

[5] J.-P. AUBIN, *Lipschitz behavior of solutions to convex minimization problems*, Math. of Operations Research 9, n° 1 (1984) 87-111.

[6] G. AUCHMUTY, *Duality for nonconvex variational principles*, J. of Differential Equations 50 (1983), 80-145.

[7] M. BOUGEARD, *Contribution à la théorie de Morse en dimension finie*, Thèse de 3ème cycle de l'Université de Paris IX, 1978.

[8] L.N. BRYSGALOVA, *Singularities of max of functions depending on parameters*, Funct. Anal. Appl. 11, 1 (1977), 59-60.

[9] L.N. BRYSGALOVA, *On max functions of families depending on parameters*, Funct. Anal. Appl. 12, 1 (1978), 66-67.

[10] H. BUSEMANN, *Convex surfaces*, Interscience Tracts in Pure and Applied Mathematics, 1958.

[11] F.H. CLARKE, *Nonsmooth analysis and optimization*, J. Wiley Interscience, 1983.

[12] V.F. DEMYANOV and L.N. POLYAKOVA, *Conditions for minimum of a quasi-differentiable function on a quasi-differentiable set*, U.S.S.R. Comput. Math. Phys. 20, (1981) 34-43.

[13] V.F. DEMYANOV and A.M. RUBINOV, *On quasi-differentiable mappings*, Math. Operationsforsch. u. Stat. ser. Optimization 14, (1983) 3-21.

[14] R.M. DUDLEY, *On second derivatives of convex functions*, Math. Scand. 41 (1977) 159-174 & 46 (1980) 61.

[15] I. EKELAND, *Legendre duality in nonconvex optimization and calculus of variations*, SIAM J. Control Optimization 15 (1977), 905-934.

[16] I. EKELAND, *Nonconvex duality*, Bull. Soc. Math. France, Mémoire 60 (1979) 45-55.

[17] I. EKELAND and J.-M. LASRY, *Problèmes variationnels non convexes en dualité*, C.R. Acad. Sc. Paris, t. 291 (1980), 493-496.

[18] R. ELLAIA, *Contribution à l'analyse et l'optimisation de différences de fonctions convexes*, Thèse de 3ème cycle de l'Université Paul Sabatier, 1984.

[19] R. ELLAIA and J.-B. HIRIART-URRUTY, *The conjugate of the difference of convex functions*, to appear in J. of Optim. Theory and Applications.

[20] D. GABAY, *Minimizing the difference of two convex functions : Part I : Algorithms based on exact regularization*, Working paper, I.N.R.I.A. (1982).

[21] P. HARTMAN, *On functions representable as a difference of convex functions*, Pacific J. Math. 9, (1959) 707-713.

[22] B. HERON and M. SERMANGE, *Non-convex methods for computing free boundary equilibria of axially symmetric plasmas*, Appl. Math. Optimization 8 (1982), 351-382.

[23] J.-B. HIRIART-URRUTY, *The approximate first-order directional derivatives for a convex function* in "Mathematical Theories of Optimization" Lecture notes in Mathematics 979, (1983) 144-177.

[24] J.-B. HIRIART-URRUTY, *A general formula on the conjugate of the difference of functions*, Séminaire d'Analyse Numérique, Université Paul Sabatier (1984).

[25] J.-B. HIRIART-URRUTY, *Miscellanies on the analysis and optimization of nonsmooth functions*, to appear.

[26] R. JANIN, *Sur la dualité et la sensibilité dans les problèmes de programmation mathematique*, Thèse Université Paris IX (1974).

[27] R. JANIN, *Sur des multiapplications qui sont des gradients généralisés*, Note aux C.R.A.S. Paris 294, (1982) 115-117.

[28] E.M. LANDIS, *On functions representable as the difference of two convex functions*, Dokl. Akad. Nauk SSSR 80, 1 (1951), 9-11.

[29] C. MALIVERT, *Méthodes de descente sur un fermé non convexe*, Bull. de la Soc. Math. de France, Mémoire n° 60, (1979) 113-124.

[30] J.N. MATHER, *Distance from a manifold in Euclidean space*, Proc. of Symp. in Pure Math., Vol. 40, Part 2 (1983), 199-216.

[31] V.I. MATOV, *Topological classification of the germs of max and minimax of the families in general position*, Russian Math. Surv. 37, 4 (1982), 167-168.

[32] V.I. MATOV, *Singularities of max functions on the manifold with boundary*, Trudy Sem. I.G. Petrovskogo, Mosk. Gos. Univ. 6 (1981), 195-222.

[33] D. MELZER, *Expressibility of piecewise linear continuous functions as a difference of two piecewise linear convex functions*, preprint Humboldt Universität Berlin (1984).

[34] F. MIGNOT, *Contrôle dans les inéquations variationnelles elliptiques*, J. of Funct. Analysis 22, (1976) 130-185.

[35] J. NOAILLES, *Une méthode d'optimisation globale en programmation non convexe*, Colloque National d'Analyse Numérique de Gouvieux, 1980.

[36] J.-P. PENOT, *Calcul sous-différentiel et optimisation*, J. of Funct. Analysis 27, (1978) 248-276.

[37] A. POMMELLET, *Analyse convexe et théorie de Morse*, Thèse de 3ème cycle de l'Université de Paris IX, 1982.

[38] B.N. PSHENICHNYI, in *Contrôle optimal et Jeux différentiels*, Cahiers de l'I.R.I.A. n° 4 (1971).

[39] B.N. PSHENICHNYI, *Necessary conditions for an extremum*, Marcel Dekker N.Y., 1971.

[40] A.W. ROBERTS and D.E. VARBERG, *Convex functions*, Academic Press, 1973.

[41] R.T. ROCKAFELLAR, *Convex analysis*, Princeton University Press, 1970.

[42] R.T. ROCKAFELLAR, *Favorable classes of Lipschitz continuous functions in subgradient optimization, in* "Progress in nondifferentiable optimization" E. Nurminskii, ed., Pergamon Press (1981).

[43] R.T. ROCKAFELLAR, *Maximal monotone relations and the second derivatives nonsmooth functions*, preprint 1984.

[44] A. SHAPIRO and Y. YOMDIN, *On functions representable as a difference of two convex functions and necessary conditions in a constrained optimization*, preprint Ben-Gurion University of the Negev (1982).

[45] A. SHAPIRO, *On functions representable as a difference of two convex functions in inequality constrained optimization*, Research report University of South Africa (1983).

[46] A. SHAPIRO, *On optimality conditions in quasidifferentiable optimization*, SIAM J. Control and Optimization 22, 4 (1984) 610-617.

[47] R. SCHNEIDER, *Boundary structure and curvature of convex bodies* in "Contributions to Geometry", J. Tölke and J.M. Wills ed., Birkhäuser Verlag, 1979.

[48] J. SPINGARN, *Submonotone subdifferentials of Lipschitz functions*, Trans. Amer. Math. Soc. 264, (1981) 77-89.

[49] J.F. TOLAND, *Duality in nonconvex optimization*, J. Math. Analysis and Applications 66 (1978), 399-415.

[50] J.F. TOLAND, *A duality principle for nonconvex optimization and the calculus of variations*, Arch. Rational Mech. Anal. 71 (1979), 41-61.

[51] J.F. TOLAND, *On subdifferential calculus and duality in nonconvex optimization*, Bull. Soc. Math. France, Mémoire 60 (1979), 177-183.

[52] H. TUY, *Global minimization of a difference of two convex functions*, to appear.

[53] J.-P. VIAL, *Strong and weak convexity of sets and functions*, Math. of Operations Research 8, 2 (1983), 231-259.

[54] Y. YOMDIN, *On functions representable as a supremum of a family of smooth functions*, SIAM J. Math. Analysis 14, 2 (1983), 239-246.

[55] Y. YOMDIN, *On functions representable as a supremum of a family of smooth functions, II*, to appear.

[56] Y. YOMDIN, *Maxima of smooth families III : Morse-Sard theorem*, preprint 1984.

[57] Y. YOMDIN, *On representability of* convex *functions as maxima of linear families*, preprint 1984.

[58] V.A. ZALGALLER, *On the representation of a function of two variables as the difference of convex functions*, Vestn. Leningrad Univ. Ser. Mat. Mekh. 18, (1963) 44-45 (in Russian).

GUIDE FOR A BIBLIOGRAPHY

Examples of d.c. optimization problems : [20], [22], [50].

Characterization of d.c. functions, decomposition of d.c. functions : [2], [3], [18, ch. 2], [21], [28], [33], [40, ch. I], [44], [58].

Analysis and optimization of lower-C^k *functions* : [18, ch. I], [26], [27], [29], [42], [48], [53], [54], [55], [56].

Singularities of max and minimax functions : [8], [9], [30], [31], [32].

Second-order derivatives of convex functions ;

pointwise derivatives : [1], [5, §2], [10], [34, §1], [43], [47, § II.4]

distributional derivatives : [14] ;

approximate derivatives : [23].

First-order necessary conditions for optimality

for directionally differentiable functions : [25], [36], [39] ;

for tangentially d.c. functions : [12], [18, ch. 2], [45], [46].

Morse theory for d.c. functions and study of their critical points : [7], [37], [56].

Duality theory for d.c. minimization problems ; formulae and applications : [6], [15], [16], [17], [19], [24], [38], [49], [50], [51].

Algorithms for minimizing d.c. functions : [20], [22], [35], [52].

FROM CONVEX TO MIXED PROGRAMMING

J. Ponstein

INTRODUCTION.

This paper is concerned with *convex programming*, *differential programming* and *mixed programming*. All spaces involved will assumed to be linear, normed and complete, so that we limit ourselves to *Banach spaces*, although in a number of cases locally convex topological vector spaces would do as well. Differentiability will be restricted to *Fréchet differentiability*. A *mixed programming problem* is an optimization problem involving two variables, x and u. Differentiability is assumed with respect to x but not with respect to u. Optimality conditions for this type of problem are a mixture of first-order conditions, in terms of the derivatives with respect to x, and global conditions with respect to u. The latter requires that certain convexity assumptions must hold.

Optimal control is an important special case of mixed programming. The abstract optimality conditions of mixed programming then lead to *minimum principles*, *Hamilton equations*, and *transversality conditions*. As will be shown another application is to *generalized linear programming*.

Much attention will be paid to *regularity conditions* that are needed in order to be able to derive necessary optimality conditions. There are two main types of such conditions, i.e. conditions similar to the well-known constraint qualification introduced by Slater, and conditions which for the linear case are based on the classic *open mapping theorem*, stating that a continuous, linear mapping from a Banach space X *onto* another Banach space Y, maps *open* sets onto *open* sets. When applied to inequality constraints, however, the latter are rather restrictive. More recent work has revealed that these can be weakened, but this requires the introduction of *convex processes*, see e.g. Rockafellar [10] and Robinson [7], a generalization of the open mapping theorem, and a generalization of a *local approximation theorem*, due to Liusternik [2]. Local approximation theorems are akin to the well-known inverse function theorems, but the former are much more suitable than the latter, as far as optimization is concerned. This was realized by Tuy [11], Robinson [9] and Nieuwenhuis [3]. Generalizations of Liusternik's result were given in Tuy [11] and Robinson [9], whereas in Nieuwenhuis [3] the argumentation was greatly streamlined, only at the cost of a slight loss in generality. The results are, however, only applicable to differentiable programming, and another generalization, given in Ponstein [6] and quite in the spirit of Nieuwenhuis [3], is required to make application to mixed programming possible. This complements results in Ioffe and Tihomirov [1], which, as far as inequality constraints are concerned, are based on Slater-type conditions.

In the final section we consider mixed minimization together with concavity assumptions.

1. A SUMMARY OF CONVEX PROGRAMMING.

The general convex optimization problem (with only a single objective, to which case we will restrict ourselves) can be written as,

$$\inf_x \{f(x) : x\in G\},$$

where x is an element of a linear space X, $f(x)\in R$, $G\subset X$, f is a convex function, and G is a convex set. In many cases G will, at least partly, be described by means of equality and inequality constraints. Since equality constraints can be described by means of inequality constraints, let us only consider inequality constraints:

$$(1.1)\qquad \inf_x \{f(x) : g(x)\leqq 0,\ x\in C\},$$

where x and f are as before, $C\subset X$, with X as before, and g(x) is an element of a Banach space Y, complete with a fixed convex cone K with apex at $0\in Y$. The meaning of the inequality in (1.1) follows from

<u>Definition 1.1.</u> If $y\in K$ then we write $y\geqq 0$, if $y\in \text{int } K$, the *interior* of K, then we write $y>0$. Further, if $-y\geqq 0$, then we write $y\leqq 0$, and if $-y>0$, then we write $y<0$. K is called the *nonnegative cone* (of Y).

The introduction of K is only meaningful, if K is a *pointed* cone, that is if

$$K\cap(-K)=\{0\}.$$

On the other hand, if, say, $Y=R^m$, it is not necessary to take for K the nonnegative quadrant of R^m. An extreme case is where

$$K=\{0\},$$

then $g(x)\leqq 0$ reduces to $g(x)=0$, so that there are two possibilities of treating equality constraints $g(x)=0$, i.e. either by setting

$$g(x)\leqq 0 \text{ and } -g(x)\leqq 0,$$

where the inequality is defined with respect to a nondegenerate K, or by letting K degenerate to $K=\{0\}$. Combinations of equality constraints and inequality constraints, can, of course, be written by means of a single inequality with respect to a suitably chosen K.

A *dual* of (1.1) can be obtained by applying Rockafellar's perturbation method, see e.g. Rockafellar [10] and the next definition.

<u>Definition 1.2.</u>

$$p(y)=\inf_x \{f(x) : g(x)\leqq y,\ x\in C\},\ y\in Y$$

$$f^d(\lambda)=\inf_y \{p(y)+\lambda y\},\ \lambda\in Y^*.$$

Here, Y* is the *conjugate space* of Y. The function p is called the *perturbation function*. It is also called the *marginal function*, apparently because, as it turns out, only the behaviour of p for small y is important. The function f^d is the *dual objective function* and the dual problem is that of finding

(1.2) $\sup_\lambda f^d(\lambda)$.

It is only necessary to consider nonnegative λ, but in order to give this a meaning, we must introduce a nonnegative cone in Y^*:

<u>Definition 1.3.</u> Given a nonnegative cone K for Y, the nonnegative cone for Y^* is defined by

$$K^*=\{\lambda : \lambda y \geqq 0 \text{ for all } y\in K\}.$$

Clearly, if $\lambda \ngeqq 0$, then $\lambda\hat{y}<0$ for some $\hat{y}\in K$, and assuming that $p(0)$ is finite, so that $g(\hat{x})\leqq 0$ for some $\hat{x}\in C$, and letting $y=g(\hat{x})+\omega\hat{y}$, where $\omega\in R$, we see that λy tends to $-\infty$ if ω tends to $+\infty$, so that $f^d(\lambda)=-\infty$. It follows that we can just as well take $\lambda\geqq 0$, but then

$$f^d(\lambda)=\inf_x \{\inf_y \{f(x)+\lambda y : y\geqq g(x)\} : x\in C\}=\inf_x \{f(x)+\lambda g(x) : x\in C\},$$

because we may then take $y=g(x)$.

One can also define f^d via the *Lagrange function*, defined by

$$L(x,\lambda)=\inf_y \{f(x)+\lambda y : y\geqq g(x)\} \text{ if } x\in C$$
$$=+\infty \qquad \text{if } x\notin C,$$

then $f_d(\lambda)=\inf_x L(x,\lambda)$.

Perturbing optimization problems by replacing $g(x)\leqq 0$ by $g(x)\leqq y$, is not the only way to construct dual problems. Another one, leading to the so-called *Fenchel duality*, see e.g. Rockafellar [10] and Ponstein [4], is where

$$\inf_x \{f(x)-g(x)\} \text{ is replaced by } \inf_x \{f(x)-g(x+y)\},$$

where $f(x)\in\{R,+\infty\}$, and $g(x)\in\{R,-\infty\}$, for all $x\in X=Y$. Letting C be the subset of X where both $f(x)$ and $g(x)$ are finite, this can also be treated by considering

$$\inf_{x_1,x_2} \{f(x_1)-g(x_2) : x_2=x_1+y,\ x_1\in C,\ x_2\in C\}.$$

Virtually any perturbation can be reduced to perturbing righthand sides of equalities or inequalities, see Ponstein [4], but in practice the reformulations may be awkward, so that the development for special cases, such as that of Fenchel duality, is worthwhile. In what follows, however, we will restrict ourselves to perturbations of righthand sides only.

Let us now turn to one of the main issues in convex programming. This is concerned with finding (sufficient) conditions under which

(1.3) $p(0)=\inf_x \{f(x) : g(x)\leqq 0,\ x\in C\}=\sup_\lambda \{f^d(\lambda) : \lambda\geqq 0\}=f^d(\lambda^o)$ for some $\lambda^o\geqq 0$.

If (1.3) is true, we say that *strong duality* holds. In passing we remark that *weak duality*, which expresses the fact that $p(0)\geqq f^d(\lambda)$ for all λ, holds trivially for the dualization procedure we are employing here. The usual conditions for strong duality are three-fold:

1) First of all, f, g and C must, of course, be convex (in fact, only p need be convex).
2) p(0) must be finite.
3) Some regularity condition must hold.

In convex programming the first condition needs no further discussion. The second one is very reasonable, so we are left with the third one, which we will consider in the next section.

2. THE TWO CLASSIC REGULARITY CONDITIONS.

A common regularity condition is that

$$g(\hat{x})<0 \text{ for some } \hat{x}\in C.$$

This is nothing other than a straight generalization of Slater's well-known constraint qualification for the case where x and g(x) are finite dimensional. This condition is theoretically easy to deal with, see e.g. Ponstein [4, Theorem 3.11.2]. Moreover, it is often easily verified if it holds, but it may not hold, because it implicitly assumes that the nonnegative cone K of Y has a nonempty interior. This is not true if, e.g. $Y=l_1$, complete with the l_1-norm, and if $K=\{y : y=(y_1,y_2,\ldots),\ y_i\geqq 0 \text{ for all } i\}$, hence if K is the nonnegative quadrant of l_1. It is also not true if $Y=R^m$ and K is the nonnegative quadrant of a linear *sub*space of Y, so that in fact certain equalities are implied by $g(x)\leqq 0$.

Another, more complicated, regularity condition is the following one, consisting of three parts. Assume that not only Y, but also X is a Banach space.

1) $g(\hat{x})\leqq 0$ for some $\hat{x}\in\text{int } C$.
   Actually, this can be weakened by replacing int C by the *relative interior* of C, which is the interior of C with respect to its *closed linear hull*.
2) f is continuous at $\hat{x}$, in case X is infinite dimensional. In fact we can do with the *upper semi-continuity* of f at $\hat{x}$ (in the sense that for all $\varepsilon>0$, there exists a δ such that $|x-\hat{x}|<\delta$ implies that $f(x)-f(\hat{x})<\varepsilon$).
3) For any (open) neighborhood U of $0\in X$, there exists a neighborhood S of the origin of Y, such that $S\subset g(\hat{x}+U)+K$. See [4, Theorem 3.11.10].

Conditions 1) and 2) seem reasonable, but condition 3) does not look so attractive. Consider, however, the case where $g(x)=b-Ax$, with $b\in Y$ and with A a continuous, linear mapping from X *onto* Y. Then, since X and Y both are assumed to be Banach spaces, it follows from the *open mapping theorem* (for which we refer the reader to the standard literature of functional analysis), that if $U\subset X$ is open, so are AU and $S=b-A(\hat{x}+U)+K=g(\hat{x}+U)+K$. The essence of the condition, therefore, is that A is a mapping *onto*, which often will be the case, in particular by taking linear subspaces of Y, if possible. Still, there may be trouble with respect to part 3) of this regularity condition, in particular in case Y is infinite dimensional. If $g(x)=b-Ax$ we can, however, relax the requirement that A is a mapping onto, and replace it by the requirement that

$$(2.1) \qquad AX+K=Y.$$

But then the open mapping theorem in its original form (i.e. a continuous, linear mapping from a Banach space onto another Banach space maps open sets onto open sets) is no longer applicable, and new tools must be made. First we need the notion of convex process, introduced by Rockafellar [10].

Definition 2.1. A *convex process* T *from* X *into* Y, where X and Y are linear spaces, is a mapping that maps the elements of X onto subsets of Y, such that $0 \in T(0)$ and such that graph $T=\{(x,y) : y \in T(x)\}$ is a convex set.

Definition 2.2. If X and Y are Banach spaces, a convex process T from X into Y is *closed* if graph T is a closed set in $X \times Y$, see Robinson [7].

Robinson [7] has shown the following generalization of the open mapping theorem.

Theorem 2.3. *Let* X *and* Y *be Banach spaces, and let* T *be a closed, convex process from* X onto Y. *Then if* $U \subset X$ *is an open set, so is* $T(U) = \cup \{T(x) : x \in U\}$.

Just as the proof of the original open mapping theorem, the proof of this theorem rests on the *Bair category theorem*, stating that if a complete, metric space is the countable union of closed subsets, at least one of these subsets contains a nonempty open set (see books on functional analysis).

Assume again that $g(x)=b-Ax$, with A as before, and also assume that the nonnegative cone of Y, that is K, is *closed*. Define T by

(2.2) $T(x)=Ax+K$.

Then it is not difficult to show that T is a closed, convex process from X into Y, and if we further assume that $AX+K=Y$, then T is convex process *onto* Y. It follows that if $U \subset X$ is open, then so is $S=b-A(\hat{x}+U)+K$, so that part 3) of the second regularity condition above is applicable. Together with [4, Theorem 3.11.10] we therefore have the following result.

Theorem 2.4. *Let* X *and* Y *be Banach spaces. Let* A *be a continuous, linear mapping from* X *into* Y. *If* p, *defined by*

$$p(y)=\inf_x \{f(x) : Ax \geqq b-y,\ x \in C\},\ y \in Y$$

is convex, if $p(0)$ *is finite, and if*

1) $A\hat{x} \geqq b$ *for some* $\hat{x} \in \text{int } C$,

2) f *is continuous at* $\hat{x}$,

3) $AX+K=Y$,

4) K *is closed,*

then, for some $\lambda^o \geqq 0$, $\lambda^o \in Y^*$,

$$p(0)=\inf_x \{f(x)+\lambda^o(b-Ax) : x \in C\}.$$

Clearly, taking K smaller results in stronger conditions regarding A. The extreme case is again $K=\{0\}$, in which case $Ax \geqq b$ becomes $Ax=b$, and $AX+K=Y$ becomes $AX=Y$, so that A must again be a mapping onto, a situation where the use of convex processes is not necessary.

3. DIFFERENTIAL PROGRAMMING.

In differential programming the convexity requirements regarding f, g and C in

(3.1) $\quad \inf_x \{f(x) : g(x) \leqq 0,\ x \in C\},\ x \in X,\ C \subset X,\ g(x) \in Y,$

are replaced by the requirement that f and g must be differentiable, and that C admits a suitable convex approximation D, see below. We will restrict ourselves to Fréchet differentiability. A function q from a Banach space X into another Banach space Z is Fréchet differentiable at $x \in X$, if there exists a continuous, linear $q'(x)$ such that

$$|q(x+h)-q(x)-q'(x)h| = o(|h|),$$

where $o(|h|)$ is some function such that

$$\lim_{h \to 0} o(|h|)/|h| = 0.$$

The function $q' : x \mapsto q'(x)$ is the *Fréchet derivative of* q. Hence, as before, we assume that X and Y are Banach spaces, and let Z be either R or Y. Further we denote the Fréchet derivatives of f and g by f' and g', respectively. The following lemma is a generalization of the mean value theorem of ordinary differentiability.

Lemma 3.1. *If* q *is Fréchet differentiable in a neighborhood of* x*, then, for* h *sufficiently small,*

$$|q(x+h)-q(x)| \leqq |h| . |q'(x+\theta h)|, \textit{ for all } \theta, 0<\theta<1.$$

In general it will not be possible to establish strong duality between (3.1) and its dual as defined by (1.2). We must satisfy ourselves with *local* optimality conditions, although these local conditions turn out to be *global* conditions of a *linearized* version of the given problem, and these global conditions are tantamount to strong duality between the linearized problem and its dual. Suppose that x^o is an optimal solution of (3.1), then the linearization we have in mind is

(3.2) $\quad \inf_x \{f'(x^o)(x-x^o) : g(x^o)+g'(x^o)(x-x^o) \leqq 0,\ x \in D\},$

with D not yet defined, although we can say already now that D must be convex and that $x^o \in D$. Perhaps the term $f(x^o)$ should have been added to the objective function, but it can, of course, just as well be omitted.

If we can show that x^o is an optimal solution of (3.2) as well, so that its infimum is equal to zero, and hence is finite, if

(3.3) $\quad g(x^o)+g'(x^o)(\hat{x}-x^o) \leqq 0$ for some $\hat{x} \in \text{int } D$,

if the nonnegative cone K of Y is closed, and if $g'(x^o)X+K=Y$, then, because $f'(x^o)$ is continuous, we can apply Theorem 2.4, and conclude that for some $\lambda^o \in Y^*$, $\lambda^o \geqq 0$,

$$0 = \inf_x \{f'(x^o)(x-x^o)+\lambda^o g(x^o)+\lambda^o g'(x^o)(x-x^o) : x \in D\}.$$

Since $x^o \in D$, it follows that $\lambda^o g(x^o) \geqq 0$, but we also have that $g(x^o) \leqq 0$ and that $\lambda^o \geqq 0$, so that

$$\lambda^o g(x^o) = 0,$$

and

$$f'(x^o)(x-x^o)+\lambda^o g'(x^o)(x-x^o)\geqq 0 \text{ if } x\in D.$$

In particular, if $x^o\in\text{int } D$, then it follows from this that

$$f'(x^o)+\lambda^o g'(x^o)=0$$

which together with $\lambda^o\geqq 0$ and $\lambda^o g(x^o)=0$ are the familiar first-order conditions for (3.1) if C is absent.

It remains to be shown that if x^o is an optimal solution of (3.1), then it is an optimal solution of (3.2) as well. In order to prove this we need a generalization of a result due to Liusternik and Sobolev [2], a result that nowadays is called a *local approximation theorem*. Such a theorem is akin to the well-known inverse function theorems, but is far more suitable with regard to optimization than are the latter. First we give a local approximation theorem in its original form, and then in its generalized form.

Lemma 3.2. *Let* q *map a Banach space* X *into another Banach space* Z, *let* q *be Fréchet differentiable in a neighborhood of some* $x\in X$, *and let* q' *be continuous at* x. *Then, if* q'(x) *is a mapping* onto, *hence if* $q'(x)X=Z$,

$$q(x)=0 \text{ and } q'(x)h=0 \text{ imply that } q(x+\tau h+o(\tau))=0 \text{ for some } o(\tau).$$

The proof of this lemma follows from the next lemma.

Lemma 3.3. *Let everything be as in Lemma 3.2, except that* Z *is equipped with a closed, nonnegative cone* K, *and that* $q'(x)X=Z$ *is replaced by* $q'(x)X+K=Z$. *Then*

$$q(x)+q'(x)\tau h\leqq 0, \text{ for } 0\leqq\tau\leqq 1, \text{ implies that } q(x+\tau h+o(\tau))\leqq 0 \text{ for some } o(\tau).$$

In essence Lemma 3.3 and its proof are due to Robinson [9]. Actually, Lemma 3.3 is slightly less general than the original result by Robinson. In the present form it has been formulated by Nieuwenhuis [3] , who gave a considerably shorter proof. This proof follows from a further generalization we will develop when considering *mixed programming*. Since the proof of that further generalization is quite similar to the proof given in Nieuwenhuis [3], we will not give a separate proof of Lemma 3.3 here.

We will use Lemma 3.3 to conclude from $g(x^o)+g'(x^o)\tau(x'-x^o)\leqq 0$ for some x', that $g(x^o+\tau(x'-x^o)+o(\tau))\leqq 0$ and that $x^o+\tau(x'-x^o)+o(\tau)\in C$. Now, x' will be such that $x(\tau)=\tau x'+(1-\tau)x^o\in D$. In other words, from the feasibililty of $x(\tau)$ with respect to the linearized problem, we will conclude the feasibility of $x(\tau)+o(\tau)$ with respect to the original problem, and draw conclusions from that. Working with inverse function theorems would be more complicated.

So, let us assume that g' is continuous at x^o, that K is closed, that $g'(x^o)X+K=Y$, and that (3.3) holds. We want to show that if x^o is an optimal solution of (3.1) then it is an optimal solution of (3.2) as well. Suppose this were not true. Then, for some $\tilde{x}\in D$,

$$f'(x^o)(\tilde{x}-x^o)<0,\ g(x^o)+g'(x^o)(\tilde{x}-x^o)\leqq 0,$$

so that for some γ, $0<\gamma<1$ and some $\varepsilon>0$,

$$f'(x^o)(\gamma(\hat{x}-x^o)+(1-\gamma)(\tilde{x}-x^o))\leqq -\varepsilon.$$

Let

$$x'=\gamma\hat{x}+(1-\gamma)\tilde{x}, \text{ and } x(\tau)=\tau x'+(1-\tau)x^o, \ 0\leqq\tau\leqq 1,$$

so that

$$x(\tau)-x^o=\tau(x'-x^o)=\tau h \text{ and } f'(x^o)(x'-x^o)\leqq-\varepsilon.$$

Since

$$g(x^o)+g'(x^o)(\hat{x}-x^o)\leqq 0,\ g(x^o)+g'(x^o)(\tilde{x}-x^o)\leqq 0, \text{ and } g(x^o)\leqq 0,$$

it follows, after multiplying these inequalities by $\tau\gamma$, $\tau(1-\gamma)$ and $1-\tau$, respectively, and adding the results, that

$$g(x^o)+g'(x^o)\tau h\leqq 0,$$

and hence, by Lemma 3.3, that

$$g(x(\tau)+o(\tau))\leqq 0.$$

Clearly,

$$x(\tau)=(1-\tau)x^o+\tau\gamma\hat{x}+\tau(1-\gamma)\tilde{x},$$

which is a convex combination of x^o, $\hat{x}$, and $\tilde{x}$, all contained in D, so that $x(\tau)\in D$. No suppose for the time being that D=C, so that C is convex. Then, because $\hat{x}\in\text{int } C$, it even follows that $x(\tau)+o(\tau)\in C$ if τ is sufficiently small. It follows that for such τ, $x(\tau)+o(\tau)$ is a feasible solution of the original problem, so that

$$f(x^o)\leqq f(x(\tau)+o(\tau)).$$

But, $f'(x^o)(x'-x^o)\leqq-\varepsilon$, or $f'(x^o)(x'-x^o+o(\tau)/\tau)\leqq-\varepsilon/2$, for $\tau>0$ and τ sufficiently small, so that

$$f'(x^o)(x(\tau)+o(\tau)-x^o)\leqq-\tau\varepsilon/2,$$

which contradicts $f(x^o)\leqq f(x(\tau)+o(\tau))$.

All we have yet to do is relax the condition that D=C is convex.

Definition 3.4. D is the *cone of internal directions* of C at x^o if D is a cone and

$$D=\{x : \tau(x+U)+(1-\tau)x^o\subset C \text{ for some neighborhood U of } 0\in X, \text{ and all } \tau, 0<\tau\leqq\tau^o, \text{ for some } \tau^o>0\}.$$

Now assume that D *is* the cone of internal directions of C at x^o, then, since $x'\in D$, $\tau(x'+U)+(1-\tau)x^o\subset C$, so that $\tau(x'+o(\tau)/\tau)+(1-\tau)x^o\in C$, for τ sufficiently small, hence that $x(\tau)+o(\tau)\in C$, and we can repeat the argument given above. Notice, however, that we do not make use of the fact that $\hat{x}$ lies in the interior of D. This is because the cone of internal directions is an open set, so that if $\hat{x}\in D$, it trivially follows that $\hat{x}\in\text{int } D$. So we have a choice: either we assume that C itself is convex and that $\hat{x}\in\text{int } C$, or we assume that $\hat{x}\in D$ and that D is the cone of internal directions of C at x^o.

From the arguments given above we derive the following theorem.

Theorem 3.5. *Let* X *and* Y *be Banach spaces, let* K *be the nonnegative cone of* Y, *and let* $f:X\to R$, *and* $g:X\to Y$ *be Fréchet differentiable. Let* x^o *be an optimal solution of*

$$\inf_x \{f(x) : g(x)\leqq 0,\ x\in C\}.$$

Then, if

1) g' *is continuous at* x^o,

2) K *is closed,*

3) $g'(x^o)X+K=Y$,

4) $g(x^o)+g'(x^o)(\hat{x}-x^o)\leqq 0$ *for some* $\hat{x}\in\text{int } D$, *where* D=C *if* C *is convex, or else* D *is the cone of internal directions of* C *at* x^o *(so that* $\hat{x}\in D$ *implies that* $\hat{x}\in\text{int } D$*)*,

for some $\lambda^o\in Y^*$, $\lambda^o\geqq 0$, $\lambda^o g(x^o)=0$,

$$f'(x^o)(x-x^o)+\lambda^o g'(x^o)(x-x^o)\geqq 0 \text{ if } x\in D.$$

As said before we can replace the condition that $\hat{x}\in\text{int } D$ by the condition that $\hat{x}\in\text{ri } D$, the relative interior of D (relative with respect to the closed linear hull of D). Further, it is sufficient to assume that f and g are differentiable in a neighborhood of x^o.

As we have seen the proof of this theorem is quite involved: it requires convex processes, a generalized open mapping theorem, and a generalized local approximation theorem. A much easier-to-prove theorem is obtained, if we replace the conditions 1) to 4) by a Slater-type condition:

$$g(x^o)+g'(x^o)(\hat{x}-x^o)<0 \text{ for some } \hat{x}\in D.$$

Moreover, f and g need only be differentiable at x^o, and in case C is not convex, D need not be the cone of internal directions of C at x^o, but can be any convex cone D satisfying

$$\text{if } x\in D, \text{ then } \tau x+(1-\tau)x^o+o(\tau)\in C \text{ for } \tau\geqq 0, \text{ and some function } o(\tau),$$

see [4, Theorem 5.4.10] for a proof.

4. MIXED PROGRAMMING, FORM OF OPTIMALITY CONDITIONS, EXAMPLES.

Mixed programming is a mixture of convex programming and differential programming:

$$(4.1)\qquad \inf_{x,u} \{f(x,u) : g(x,u)\leqq 0,\ x\in C,\ u\in V\}.$$

Here, x is an element of a space X, u is an element of a space U (notice that we used the symbol U before to denote a neighborhood in X), $C\subset X$, $V\subset U$, f(x) is a real number, g(x,u) is an element of a space Y with nonnegative cone K, determining the inequality in $g(x,u)\leqq 0$, and f and g are assumed to be differentiable. We assume that U is a *topological vector space*, and, as before, that X and Y are Banach spaces, and that f and g are Fréchet differentiable with respect to x. The derivatives will be denoted by f'_x and g'_x, respectively. Again the aim is to find necessary optimality conditions. These should have a *local* character as far as x is concerned, and have a *global* character as far as u is concerned. First of all, let us see how these conditions will look like.

Ideally, that is if everything is convex, and if suitable regularity conditions are satisfied, we may expect that for some $\lambda^o \in Y^*$, $\lambda^o \geqq 0$, we have that $\lambda^o g(x^o,u^o)=0$, and that

$$f(x^o,u^o)=\inf_{x,u}\ \{f(x,u)+\lambda^o g(x,u)\ :\ x\in C,\ u\in V\}$$
$$=\inf_x\ \{f(x,u^o)+\lambda^o g(x,u^o)\ :\ x\in C\}$$
$$=\inf_u\ \{f(x^o,u)+\lambda^o g(x^o,u)\ :\ u\in V\}.$$

The first equality here would follow from convex programming, as summarized in Section 1, and the other two equalities would follow from the fact that (x^o,u^o) is an optimal solution of $\inf_{x,u}\ \{f(x,u)+\lambda^o g(x,u)\ :\ x\in C,\ u\in V\}$ as well. So, the conditions we may expect to hold under certain assumptions are

(4.2a) $\quad f'_x(x^o,u^o)(x-x^o)+\lambda^o g'_x(x^o,u^o)(x-x^o)\geqq 0$ if $x\in C$

and

(4.2b) $\quad f(x^o,u)+\lambda^o g(x^o,u)\geqq f(x^o,u^o)=f(x^o,u^o)+\lambda^o g(x^o,u^o)$ if $u\in V$.

Inequality (4.2a) follows from the discussions in Section 3, and inequality (4.2b) follows directly from the third equality above.

For reasons to be discussed when considering *optimal control* as an example below, conditions (4.2) may be said to express a *minimum principle*.

Clearly, (4.2a) has the local character we got acquainted with in Section 3, and (4.2b) has a global character, similar to the optimality conditions of convex programming, summarized in Section 1. It will also be clear that (4.2b) will only hold if certain convexity requirements are satisfied. Somewhat too strong convexity conditions are the following ones,

C is a convex set, and also S defined by

(4.3) $\quad S=\{s\ :\ s=(s_1,s_2),\ s_1\geqq f(x^o,u),\ s_2\geqq g(x^o,u)$ for some $u\in V\}$

is a convex set.

As in optimal control, we can reduce (4.1) to a kind of *Mayer form*, by adding a real variable x', replacing the objective function by x', and adding the constraint

$$f(x,u)-x'\leqq 0.$$

In what follows, we will, however, stick to (4.1), which, although theoretically less convenient, can be applied more directly. Before going into details, however, let us first show what follows from (4.2) for a few examples.

<u>A simple optimal control problem.</u>

Consider

$$\inf_{x,u}\ \{\phi_1(x(1))+\int_0^1 \phi_2(x(t),u(t),t)dt\ :\ x(t)=\xi_0+\int_0^t \psi(x(\tau),u(\tau),\tau)d\tau,\ 0\leqq t\leqq 1,\ u\in V\}$$

where $x:t\mapsto x(t)\in R^n$, $u:t\mapsto u(t)\in R^m$, $\phi_1:\xi\mapsto\phi_1(\xi)\in R$, $\phi_2:(\xi,\mu,\tau)\mapsto\phi_2(\xi,\mu,\tau)\in R$, $\psi:(\xi,\mu,\tau)\mapsto\psi(\xi,\mu,\tau)\in R^n$, $\xi_0\in R^n$, $V\subset U$. Given are ϕ_1, ϕ_2, ψ, ξ_0 and V, and it is assumed that ϕ_1, ϕ_2 and ψ are differentiable with respect to ξ. Then

$$f(x,u)=\phi_1(x(1))+\int_0^1 \phi_2(x(t),u(t),t)dt$$

$$g(x,u):t \mapsto g(x,u)(t)$$

with

$$g(x,u)(t)=x(t)-\xi_0-\int_0^t \psi(x(\tau),u(\tau),\tau)d\tau,\quad 0\leq t\leq 1.$$

The nonnegative cone K of Y is degenerate, that is, K={0}, and the condition $x\in C$ is absent, so that (4.2a) implies that

(4.4) $\quad f'_x(x^o,u^o)h+\lambda^o g'_x(x^o,u^o)h=0$ for all $h\in X$.

Suppose that Y consists of functions $y:t\mapsto y(t)\in R^n$, such that $\lambda y=\int_0^1 \lambda(t)y(t)dt$, $\lambda(t)\in R^n$. Letting Λ be defined by $\Lambda(t)=\int_t^1 \lambda(\tau)d\tau$, we can evaluate $\lambda g(x,u)$, and applying partial integration, we get, after a straightforward computation that

$$\lambda g(x,u)=-\int_0^1 \Lambda'(t)x(t)dt-\Lambda(0)\xi_0-\int_0^1 \Lambda(t)\psi(x(t),u(t),t)dt$$

Here $\Lambda'(t)$ is the derivative of $\Lambda(t)$, and we used the fact that $\Lambda(1)=0$.

The next thing to do is to compute $f'_x(x,u)$ and $\lambda g'_x(x,u)$, or rather $f'_x(x,u)h$ and $\lambda g'_x(x,u)h$. The results are

$$f'_x(x,u)h=\phi'_1(x(1))h(1)+\int_0^1 \phi'_{2,x}(x(t),u(t),t)h(t)dt$$

and

$$\lambda g'_x(x,u)h=-\int_0^1 \Lambda'(t)h(t)dt-\int_0^1 \Lambda(t)\psi'_x(x(t),u(t),t)h(t)dt.$$

From (4.4) we get from this that

(4.5) $\quad \phi'_1(x^o(1))=0,\quad \phi'_{2,x}(x^o(t),u^o(t),t)-\Lambda^{o\prime}(t)-\Lambda^o(t)\psi'_x(x^o(t),u^o(t),t)=0,\ 0\leq t\leq 1.$

Defining the *Hamiltonian* by

(4.6) $\quad H(\xi,\mu,\tau,\sigma)=\phi_2(\xi,\mu,\tau)-\sigma\psi(\xi,\mu,\tau),\ \xi\in R^n,\ \mu\in R^m,\ \sigma\in R^n,\ \tau\in R,$

the second equation of (4.5) becomes one of the *Hamilton equations*,

(4.7) $\quad \Lambda^{o\prime}(t)=H'_\xi(x^o(t),u^o(t),t,\Lambda^o(t)),\ 0\leq t\leq 1.$

Via (4.4) we have thus found the result from (4.2a). From (4.2b) and (4.6) we obtain

(4.8) $\quad \int_0^1 H(x^o(t),u^o(t),t,\Lambda^o(t))dt\leq\int_0^1 H(x^o(t),u(t),t,\Lambda^o(t))dt$ if $u\in V$.

If V is such that if $u\in V$ also $\hat{u}\in V$, where

$$\hat{u}(t)=u^o(t),\ t\notin W,\ \text{and}\ \hat{u}(t)=u(t),\ t\in W,\ \text{for any subinterval } W \text{ of } [0,1],$$

then (4.8) would imply

(4.9) $\quad H(x^o(t),u^o(t),t,\Lambda^o(t))\leq H(x^o(t),u(t),t,\Lambda^o(t))$, if $u\in V$ and if $0\leq t\leq 1$.

which is what usually is called the *minimum principle* of the control problem we started from. This may be used to justify calling (4.2) the minimum principle of the abstract problem (4.1). If V=U and if H would be differentiable with respect to u, then from (4.9) the remaining *Hamilton equations* would follow:

(4.10) $H'_\mu(x^o(t),u^o(t),t,\Lambda^o(t))=0,\ 0\leqq t\leqq 1.$

Finally, notice that Λ is the *adjoint variable*, satisfying the differential equation (4.7), also called *adjoint equation*, and that Λ satisfies the *end condition* $\Lambda(1)=0$, whereas x has to satisfy the *initial condition* $x(0)=\xi_0$.

A variation of the optimal control problem.

A variation is obtained if we delete the term $\phi_1(x(1))$ from the objective function, replace ϕ_2 by ϕ, ψ by ψ_1 and add the constraint

$$\psi_2(x(1))\leqq 0, \text{ with } \psi_2(x(1)) \text{ in a Banach space } Y_2.$$

Then an entirely similar computation would give, for some $\lambda_2^o\in Y_2^*$, $\lambda_2^o\geqq 0$,

$$\Lambda_1^{o\prime}(t)=\lambda_2^o\psi_2'(x^o(1))+H_\xi'(x^o(t),u^o(t),t,\Lambda_1^o(t)),\ 0\leqq t\leqq 1,\ \Lambda_1^o(1)=0$$

$$\int_0^1 H(x^o(t),u^o(t),t,\Lambda_1^o(t))dt \leqq \int_0^1 H(x^o(t),u(t),t,\Lambda_1^o(t))dt$$

where now

$$H(\xi,\mu,\tau,\sigma)=\phi(\xi,\mu,\tau)-\sigma\psi_1(\xi,\mu,\tau).$$

In addition we would have the following *transversality conditions*,

$$\lambda_2^o\psi_2(x^o(1))=0,$$

resulting from the abstract equation $\lambda^o g(x^o,u^o)=0$.

Generalized linear programming.

Let $x=(x_1,\dots,x_n)$, $x_i\in R$, and replace u by the pair (u,v), with $u=(u_1,\dots,u_n)$, $u_i\in R$, $v=(v_1,\dots,v_n)$. Let v_i and b be m-dimensional vectors.

Consider the following *generalized linear programming* problem,

$$\inf_{x,u,v} \{\Sigma u_i x_i \ :\ \Sigma v_i x_i=b,\ x_i\geqq 0,\ (u_i,v_i)\in V_i,\ i=1,\dots,n\},$$

so that $V=V_1\times\dots\times V_n$ and C is absent. The abstract minimum principle (4.2) is applicable to this problem if $V_1,\dots,V_n$ are convex sets. Then, letting λ be the multiplier for $b-\Sigma v_i x_i=0$, and μ that for $-x_i\leqq 0$, we get from (4.2), after some elementary computations,

$$u_i^o-\lambda^o v_i-\mu_i^o=0,\ i=1,\dots,n$$

and

(4.11) $u_i^o x_i^o-\lambda^o v_i^o x_i^o\leqq u_i x_i^o-\lambda^o v_i x_i^o$ if $(u_i,v_i)\in V_i$, $i=1,\dots,n.$

Moreover, we have that $\mu^o\geqq 0$, and that $\mu_i^o x_i^o=0$, $i=1,\dots,n$. It follows that the left-hand side of the inequality (4.11) is equal to zero, so that it is equivalent to

$$x_i^o \inf_{u,v} \{u_i-\lambda^o v_i \ :\ (u_i,v_i)\in V_i\}=0,\ i=1,\dots,n.$$

Depending on the structure of the V_i the problem can be solved by a suitable numerical procedure. The optimal x^o is, of course, also an optimal solution of the following ordinary linear programming problem, (whose dual is solved by λ^o),

$$\inf_x \{\Sigma u_i^o x_i : \Sigma v_i^o x_i = b,\ x_i \geq 0,\ i=1,\ldots,n\}.$$

The usual approach to generalized linear programming is by applying the *decomposition principle* of linear programming. Although the present approach does not add anything new to what we already know about generalized linear programming, it brings us quite quickly to the required optimality conditions, without having to resort to any decomposition technique.

A case where the abstract minimum principle does not work.

Consider

$$\inf_{x,u} \{2x_1^2 - x_2 : u - x_1 = 0,\ u^2 - x_2 = 0,\ 0 \leq u \leq 1\}.$$

Trivially, it follows that $x_1^o = x_2^o = u^o = 0$, but the minimum principle does not work. If it did, we would have, with $\lambda^o = (\lambda_1^o, \lambda_2^o)$,

$$4x_1^o - \lambda_1^o = 0 \text{ and } -1 - \lambda_2^o = 0$$

together with

$$\inf_u \{\lambda_1^o u + \lambda_2^o u^2 : 0 \leq u \leq 1\} = 0,$$

so that $\lambda_1^o = 0$, $\lambda_2^o = -1$, and hence $-1=0$, contradiction. The reason for the failure is that the set S is not convex.

A similar case where the abstract minimum principle does work.

Consider the next problem, which looks very similar to the previous one,

$$\inf_{x,u} \{2x_1^2 - x_2 : x_1 = \int_0^1 u(t)dt,\ x_2 = \int_0^1 u^2(t)dt,\ u \in V\},$$

with V the set of all piece-wise continuous functions $u: t \mapsto u(t)$, $0 \leq u(t) \leq 1$. Now S is convex, and the minimum principle gives again $\lambda_1^o = 4x_1^o$ and $\lambda_2^o = -1$, and further

$$\inf_u \{\int_0^1 [4x_1^o u(t) - u^2(t)]dt,\ u \in V\} = 0.$$

Since the integrand here is concave, it easily follows that u^o must be *bang-bang*, that is either $u(t)=0$ or $u(t)=1$, for all t, $0 \leq t \leq 1$. If μ is the measure of the set of all t's where $u^o(t)=1$, then it follows that

$$\inf_\mu \{4x_1^o \mu - \mu : 0 \leq \mu \leq 1\} = 0 \text{ and } x_1^o = \int_0^1 u^o(t)dt = \mu,$$

so that $\mu = 1/4 = x_1^o = x_2^o$, and the original infimum is equal to $-1/8$.

Notice, that for showing that S is convex, the convexity of the interval [0,1], is irrelevant, as we could just as well have required that, say, either $0 \leq u(t) \leq 1/4$, or $3/4 \leq u(t) \leq 1$, instead of $0 \leq u(t) \leq 1$.

Also notice that if the constraints are $x_1 = (u_1 + \ldots + u_n)/n$, $x_2 = (u_1^2 + \ldots + u_n^2)/n$, $0 \leq u_i \leq 1$, i=1.,...,n, the minimum principle would not work, except if n=4t for integer t. Clearly, the fact that S is convex if we work with integrals, bears on *Liapounoff's convexity theorem,* stating that if $\mu_1,\ldots,\mu_n$ are finite positive *nonatomic* measures on a measure space Z, then the set of points in R^n of the form

$$(\mu_1(E),\ldots,\mu_n(E)),$$

with E ranging over all measurable subsets of Z, is closed and *convex*. Apparently, the nonatomicness of these measures is crucial.

5. REGULARITY CONDITIONS FOR MIXED PROGRAMMING.

Let us return to the abstract mixed programming problem

$$\text{(5.1)} \qquad \inf_{x,u} \{f(x,u) : g(x,u) \leqq 0,\ x \in C,\ u \in V\}.$$

Assume that (x^o,u^o) is an optimal solution of this problem, so that the infimum is finite, and assume that both C and

$$\text{(5.2)} \qquad S=\{s : s=(s_1,s_2),\ s_1 \in R,\ s_2 \in Y,\ s_1 \geqq f(x^o,u),\ s_2 \geqq g(x^o,u) \text{ for some } u \in V\}$$

are convex sets (later on we will relax these conditions). The inequality involving s_1 is, of course, determined by the complete ordering of R, that involving s_2 is determined by the partial ordering given by the nonnegative cone K. It remains to specify regularity conditions required in order to show that for some λ^o,

$$\text{(5.3a)} \qquad f'_x(x^o,u^o)(x-x^o)+\lambda^o g'(x^o,u^o)(x-x^o) \geqq 0 \text{ if } x \in C$$

$$\text{(5.3b)} \qquad f(x^o,u)+\lambda^o g(x^o,u) \geqq f(x^o,u^o) \text{ if } u \in V$$

$$\text{(5.3c)} \qquad \lambda^o \geqq 0,\ \lambda^o g(x^o,u^o)=0.$$

Slater-type regularity conditions are considered in Ioffe and Tihomirov [1] and Ponstein [5]. In [1] the convexity conditions are weaker than in [5], where they are the same as in (5.2), but whereas in [1] equality constraints may be quite arbitrary, the number of *scalar* inequalities must be *finite*, which may be related to the fact that *Liapounoff's convexity theorem* (see end of Section 4) is invoked. Also a condition like $x \in C$ is not taken up. In case Slater-type conditions are not applicable, neither [1], nor [5] can be used, and additions to the theory have to be developed. This will be the subject of the present and the next section, which give a worked-out version of Ponstein [6]. One of the tools is a further generalization of the local approximation theorem (see Lemma 3.3). Its proof closely follows that of Lemma 3.3, as given in Nieuwenhuis [3].

Let us start with a number of preliminary definitions and lemma's. Notice that in what follows we consider convex processes from Y into X, not from X into Y.

Definition 5.1. (Robinson [7]). A convex process from a Banach space Y into a Banach X is called *normed* if $|T|$, defined by

$$|T|=\sup_{|y|\leqq 1} \inf_{x \in T(y)} |x|$$

is a finite number. Then $|T|$ is called the *norm* of T.

<u>Lemma 5.2.</u> (Robinson [7, Corollary of Theorem 2], 'generalized closed graph theorem'). *If* T *is a closed convex process from a Banach space* Y <u>onto</u> *a Banach space* X, *and if* $T(y)\neq\emptyset$ *for all* $y\in Y$, *then* $|T|$ *is finite, hence* T *is normed.* (Here, 'onto' means that $\bigcup_y T(y)=X$.)

<u>Definition 5.3.</u> Let P and Q be subsets of a Banach space Y, then, by definition,

$$P-Q=\{\, y : y=p-q \text{ for some } p\in P \text{ and some } q\in Q\}$$

$$d(y,P)=\inf_p \{|p-y| : p\in P\}$$

$$d(P,Q)=\sup_p \{d(p,Q) : p\in P\}$$

$$\rho(P,Q)=\max\{d(P,Q),d(Q,P)\},$$

so that $\rho(P,Q)$ is the *Hausdorff distance* between P and Q.

<u>Lemma 5.4.</u> (Robinson [7, Theorem 6]).
If T *is a normed convex process from a Banach space* Y *into a Banach space* X, *and if* P *and* Q *are nonempty subsets of* Y *such that* $T(P)\neq\emptyset$ *and* $T(Q)\neq\emptyset$ *then*
(a) if $T(y)\neq\emptyset$ *for all* $y\in Q-P$, *then* $d(T(P),T(Q))\leq|T|.d(P,Q)$,
(b) if $T(y)\neq\emptyset$ *for all* $y\in(Q-P)\cup(P-Q)$, *then* $\rho(T(P),T(Q))\leq|T|.\rho(P,Q)$.
(Here, T(P) is, of course, the union of all T(p), $p\in P$, and similarly for T(Q).)

We will only use part (b) of this lemma; for $P=\{y_1\}$ and $Q=\{y_2\}$.

<u>Lemma 5.5.</u> (Robinson [8, Lemma]).
Let (X,ρ) *be a complete metric space, and let* F *be any function from* X *into* 2^X, *such that there exist* $\hat{x}\in X$, *and real numbers* α *and* r, $0<\alpha<1$, $r\geq0$, *for which*
(a) for some $\varepsilon>0$ *and all* x_1, x_2 *in the closed ball* $B(\hat{x},r+\varepsilon)$ *of radius* $r+\varepsilon$ *around* $\hat{x}$, $F(x_1)$ *and* $F(x_2)$ *are nonempty and closed, and* $\rho(F(x_1),F(x_2))\leq\alpha\rho(x_1,x_2)$, *and*
(b) $d(\hat{x},F(\hat{x}))\leq(1-\alpha)r$.
Then there exists an $x_\infty\in B(\hat{x},r+\varepsilon)$ *such that* $x_\infty\in F(x_\infty)$ (so that x_∞ is a fixed point of F), *and* $\rho(\hat{x},x_\infty)\leq(1-\alpha)^{-1}d(\hat{x},F(\hat{x}))+\varepsilon$.

After these preparations we can now prove

<u>Theorem 5.6.</u> ('generalized local approximation theorem for the mixed case').
Let X *and* Y *be Banach spaces, and let* U *be a topological vector space. Assume that the nonnegative cone* K *of* Y *is closed, and that*
(a) $g:X\times U\to Y$ *is Fréchet differentiable with respect to* x *in a neighborhood of* $(x^o,u^o)\in X\times U$,
(b) g'_x *is continuous at* (x^o,u^o), *and*
(c) $g'_x(x^o,u^o)X+K=Y$.
Then $g(x^o,u(\tau))+g'_x(x^o,u^o)\tau h\leq0$, $0\leq\tau\leq1$, *and* $\lim_{\tau\downarrow0} u(\tau)=u(0)=u^o$, *imply that*

$$g(x^o+\tau h+o(\tau),u(\tau))\leq0 \textit{ for some } o(\tau) \textit{ with } \lim_{\tau\downarrow0}|o(\tau)|/\tau=0.$$

<u>Proof.</u> The proof is similar to the proof of the main theorem in Nieuwenhuis [3]. In the proof we use $o(\tau)$ for different purposes, but in all cases it is an order-of

function, whose norm tends to zero faster than does τ itself (for $\tau \geqq 0$). The result is trivial if $h=0$, hence let $h \neq 0$. Further, let $\tau>0$.

Part A. First we show that

$$g(x^o+\tau h,u(\tau))=g(x^o,u(\tau))+g_x'(x^o,u^o)\tau h+o(\tau).$$

To see this, put

$$r(\tau)=g(x^o+\tau h,u(\tau))-g(x^o,u(\tau))-g_x'(x^o,u^o)\tau h=q(x^o+\tau h)-q(x^o)$$

where

$$q(x^o+\tau h)=g(x^o+\tau h,u(\tau))-g_x'(x^o,u^o)(x^o+\tau h).$$

By the generalized mean value theorem, see Lemma 3.1,

$$|q(x^o+\tau h)-q(x^o)|\leqq|\tau h|.|q'(x^o+\theta\tau h)| \text{ for all } \theta,\ 0<\theta<1,$$

or

$$|r(\tau)/\tau|\leqq|h|.|g_x'(x^o+\theta\tau h,u(\tau))-g_x'(x^o,u^o)| \text{ for all } \theta,\ 0<\theta<1.$$

From the continuity of g_x' at (x^o,u^o) and $\lim u(\tau)=u(0)=u^o$ it follows from this that $r(\tau)=o(\tau)$, from which the desired result immediately follows.

Part B. Next we define $T:Y\to 2^X$ and $F\ :X\to 2^X$ by

$$T(y)=\{x\ :\ g_x'(x^o,u^o)x+y\leqq 0\}$$

and

$$F_\tau(x')=\{x\ :\ g(x^o+\tau h+x',u(\tau))+g_x'(x^o,u^o)(x-x')\leqq 0\}.$$

Then if we let $y'=g(x^o+\tau h+x',u(\tau))-g_x'(x^o,u^o)x'$ it follows that $F_\tau(x')=T(y')$.
Further we have that $F_\tau(x')$ is nonempty and closed for any $x'\in X$, because $g_x'(x^o,u^o)X+K=Y$, so that for $y=g(x^o+\tau h+x',u(\tau))$, there exist $x-x'\in X$ and $k\in K$, such that $g_x'(x^o,u^o)(x-x')+k=-y$, or $y+g_x'(x^o,u^o)(x-x')\leqq 0$, so that $x\in F_\tau(x')$ and $F_\tau(x')\neq\emptyset$. $F_\tau(x')$ is closed because K is closed.
Clearly, T is a convex process from Y into X. T is closed, because K is closed, and T is onto, because $g_x'(x^o,u^o)X+K=Y$. Hence, by Lemma 5.2, T is normed, so that $|T|$ is finite.

Part C. Put

$$\sigma(\tau)=4\{|T|.|r(\tau)|+\tau^2.|h|^2\}.$$

Then, by Part A, $\sigma(\tau)=o(\tau)$, and we trivially have that $\sigma(\tau)>0$ if $\tau>0$. Further we have:

$$d(0,F_\tau(0))=\inf_x\ \{|x|\ :\ x\in F_\tau(0)\}=\inf_x\ \{|x|\ :\ g(x^o+\tau h,u(\tau))+g_x'(x^o,u^o)x\leqq 0\}$$
$$\leqq\inf_x\ \{|x|\ :\ g(x^o+\tau h,u(\tau))+g_x'(x^o,u^o)x+k\leqq 0\} \text{ for any } k\in K.$$

In particular, take $k=-g(x^o,u(\tau))-g_x'(x^o,u^o)\tau h$, which if τ is small enough, indeed is an element of K. By the definition of $r(\tau)$, see Part A, it follows from this that

$$d(0,F_\tau(0))\leqq\inf_x\ \{|x|\ :\ r(\tau)+g_x'(x^o,u^o)x\leqq 0\}.$$

But

$$|T| = \sup_{|y|\leq 1} \inf_x \{|x| : g_x'(x^o,u^o)x+y\leq 0\}$$

$$\geq \inf_x \{|x| : g_x'(x^o,u^o)x+r(\tau)/|r(\tau)|\leq 0\}, \text{ if } r(\tau)\neq 0,$$

hence

$$|r(\tau)|.|T|\geq \inf_x \{|x| : g_x'(x^o,u^o)x+r(\tau)\leq 0\} \leq d(0,F_\tau(0)), \text{ also if } r(\tau)=0,$$

or

$$d(0,F_\tau(0))\leq |r(\tau)|.|T|\leq \tfrac{1}{4}\sigma(\tau).$$

Let x_1 and x_2 be any two elements contained in the closed ball $B(0,\sigma(\tau))$ around $\hat{x}=0$ with radius $\sigma(\tau)$, and let $y_i=g(x^o+\tau h+x_i,u(\tau))-g_x'(x^o,u^o)x_i$, $i=1,2$, so that according to Part B, $F_\tau(x_i)=T(y_i)$, $i=1,2$. Then it follows from Definition 5.3 and Lemma 5.4 that

$$\rho(F_\tau(x_1),F_\tau(x_2))=\rho(T(y_1),T(y_2))\leq |T|.\rho(y_1,y_2)=|T|.|y_1-y_2|.$$

By the generalized mean value theorem, Lemma 3.1 (see also Part A),

$$|y_1-y_2|\leq |x_1-x_2|.|g_x'(x^o+\tau h+x_2+\theta(x_2-x_1),u(\tau))-g_x'(x^o,u^o)|\leq \tfrac{1}{2}.|x_1-x_2|/|T|$$

if τ is small enough and if $|T|\neq 0$. Hence

$$\rho(F_\tau(x_1),F_\tau(x_2))\leq \tfrac{1}{2}.|x_1-x_2|,$$

also if $|T|=0$.

We can now apply Lemma 5.5, and taking $\alpha=\frac{1}{2}$, $r=\varepsilon=\frac{1}{2}\sigma(\tau)$, so that $r+\varepsilon=\sigma(\tau)$, and $\hat{x}=0$, we get,

$$x_\infty\in F_\tau(x_\infty) \text{ for some } x_\infty\in B(0,\sigma(\tau)),$$

and since $x_\infty=o(\tau)$, it follows from the definition of $F_\tau(x)$ that

$g(x^o+\tau h+o(\tau),u(\tau))\leq 0$. <u>End of proof</u>.

6. DERIVING MINIMUM PRINCIPLES FOR MIXED PROGRAMMING.

In this section we will give a proof (based on Theorem 5.6) for a minimum principle for mixed programming, if Slater-type regularity conditions cannot be applied. Further, we will discuss what happens if Slater-type conditions are applicable, but we will give no (complete) proofs, as they can be found in Ioffe and Tihomirov [1] and Ponstein [5] or elsewhere.

<u>Theorem 6.1.</u> *Let* X *and* Y *be Banach spaces, and let* U *be a topological vector space. Let* K *be the nonnegative cone of* Y, *let* $f:X\times U\to R$ *and* $g:X\times U\to Y$ *be given, let* f *and* g *be Fréchet differentiable with respect to* x *in a neighborhood of a point* $(x^o,u^o)\in X\times U$, *and let* $C\subset X$ *and* $V\subset U$ *be given.*
Assume that (x^o,u^o) *is an optimal solution of*

$$\inf_{x,u} \{f(x,u) : g(x,u)\leq 0,\ x\in C,\ u\in V\}$$

and that C *and* S *defined by*

$S=\{s : s=(s_1,s_2),\ s_1\in R,\ s_2\in Y,\ s_1\geqq f(x^o,u),\ s_2\geqq g(x^o,u)$ for some $u\in V\}$

are convex sets. Moreover, assume that, given $\tilde{u}\in V$, $0\leqq\tau\leqq 1$, *not only*

$$f(x^o,u(\tau))\leqq\tau f(x^o,\tilde{u})+(1-\tau)f(x^o,u^o) \text{ and } g(x^o,u(\tau))\leqq\tau g(x^o,\tilde{u})+(1-\tau)g(x^o,u^o)$$

for some $u(\tau)\in V$ (as follows from the convexity of S), *but also that*

$$\lim_{\tau\downarrow 0} u(\tau)=u(0)=u^o.$$

Then, if

1) g'_x *is continuous at* (x^o,u^o),

2) K *is closed,*

3) $g'_x(x^o,u^o)X+K=Y$,

4) $g(x^o,\hat{u})+g'_x(x^o,u^o)(\hat{x}-x^o)\leqq 0$ *for some* $(\hat{x},\hat{u})\in(\text{int } C)\times V$,

there exists a $\lambda^o\in Y^*$, $\lambda^o\geqq 0$, $\lambda^o g(x^o,u^o)=0$, *such that*

$$\left.\begin{array}{ll} f'_x(x^o,u^o)(x-x^o)+\lambda^o g'_x(x^o,u^o)(x-x^o)\geqq 0 & \text{if } x\in C \\ f(x^o,u)+\lambda^o g(x^o,u)\geqq f(x^o,u^o) & \text{if } u\in V \end{array}\right\} \text{ minimum principle.}$$

Proof. The structure of the proof is quite similar to that of the proof of Theorem 3.5, which in fact is a special case of the present theorem. In Part A of the proof we show that (x^o,u^o) is an optimal solution of a linearized version of the given problem as well, thereby using Theorem 5.6. In Part B we derive the minimum principle from this, by applying standard convex programming theory.

Part A. We claim that (x^o,u^o) is also an optimal solution of

$$\inf_{x,u}\ \{f(x^o,u)+f'_x(x^o,u^o)(x-x^o) : g(x^o,u)+g'_x(x^o,u^o)(x-x^o)\leqq 0,\ x\in C,\ u\in V\}.$$

Since (x^o,u^o) is a feasible solution of this problem, its infimum is at most equal to $f(x^o,u^o)$, and if (x^o,u^o) is not optimal, then this infimum is smaller than $f(x^o,u^o)$. Suppose the latter were true, then we arrive at a contradiction, as follows. For some $(\tilde{x},\tilde{u})\in C\times V$,

$$f(x^o,\tilde{u})+f'_x(x^o,u^o)(\tilde{x}-x^o)<f(x^o,u^o) \text{ and } g(x^o,\tilde{u})+g'_x(x^o,u^o)(\tilde{x}-x^o)\leqq 0.$$

Take $0<\gamma<1$, and take γ small enough such that for some $\varepsilon>0$,

$$\gamma f(x^o,\hat{u})+(1-\gamma)f(x^o,\tilde{u})+f'_x(x^o,u^o)(\gamma(\hat{x}-x^o)+(1-\gamma)(\tilde{x}-x^o))\leqq f(x^o,u^o)-\varepsilon.$$

Let

$$x'=\gamma\hat{x}+(1-\gamma)\tilde{x} \text{ and } x(\tau)=\tau x'+(1-\tau)x^o,\ 0\leqq\tau\leqq 1,$$

so that

$$x(\tau)-x^o=\tau(x'-x^o)=\tau h,$$

and

$$\gamma f(x^o,\hat{u})+(1-\gamma)f(x^o,\tilde{u})+f'_x(x^o,u^o)(x'-x^o)\leqq f(x^o,u^o)-\varepsilon.$$

From the assumptions it follows that for some $u'\in V$ and some $u(\tau)\in V$, such that $\lim_{\tau\downarrow 0} u(\tau)=u(0)=u^o$,

$$f(x^o,u')\leqq\gamma f(x^o,\hat{u})+(1-\gamma)f(x^o,\tilde{u})$$

$$g(x^o,u')\leqq\gamma g(x^o,\hat{u})+(1-\gamma)g(x^o,\tilde{u})$$

and

$$f(x^o,u(\tau))\leqq\tau f(x^o,u')+(1-\tau)f(x^o,u^o)$$

$$g(x^o,u(\tau))\leqq\tau g(x^o,u')+(1-\tau)g(x^o,u^o).$$

From this and the assumptions regarding $\hat{u}$ and $\tilde{u}$ it follows by an elementary calculation that

$$g(x^o,u(\tau))+g'_x(x^o,u^o)\tau h\leqq(1-\tau)g(x^o,u^o)\leqq 0,$$

so that, by Theorem 5.6,

$$g(x^o+\tau h+o(\tau),u(\tau))\leqq 0$$

or

$$g(x(\tau)+o(\tau),u(\tau))\leqq 0 \text{ for some } o(\tau),$$

so that

$$f(x(\tau)+o(\tau),u(\tau))\geqq f(x^o,u^o),$$

because $(x(\tau)+o(\tau),u(\tau))$ is a feasible solution of the given problem if τ is small enough. The latter follows from the fact that $x(\tau)$ is a convex combination of x^o, $\hat{x}$ and $\tilde{x}$, namely $(1-\tau)x^o+\tau\gamma\hat{x}+\tau(1-\gamma)\tilde{x}$, and because $\hat{x}\in\text{int } C$.
But, from the assumptions regarding ε we have, again by an elementary calculation, that

$$f(x^o,u(\tau))+f'_x(x^o,u^o)(x(\tau)-x^o)\leqq f(x^o,u^o)-\tau\varepsilon$$

or, if τ is small enough,

$$f(x^o,u(\tau))+f'_x(x^o,u^o)(x(\tau)+o(\tau)-x^o)\leqq f(x^o,u^o)-\tau\varepsilon/2,$$

which contradicts $f(x(\tau)+o(\tau),u(\tau))\geqq f(x^o,u^o)$, because $u(\tau)$ tends to u^o if τ tends to 0.
Part B. From Part A we see that (0,0) is a boundary point of the set

$$S'=\{(s_1,s_2) : s_1\in R,\ s_2\in Y,\ s_1\geqq f(x^o,u)+f'_x(x^o,u^o)(x-x^o),$$
$$s_2\geqq g(x^o,u)+g'_x(x^o,u^o)(x-x^o),\ \text{for some } (x,u)\in C\times V\}.$$

Moreover, since S is convex, so is S', and the interior of S' is nonempty. To see the latter, consider $T:X\to 2^Y$, defined by

$$T(x)=g'_x(x^o,u^o)x+K.$$

Then T is a closed convex process from X onto Y, since $g'_x(x^o,u^o)X+K=Y$.
Since $\hat{x}\in\text{int } C$, there is a neighborhood Q of $0\in X$, such that $\hat{x}+Q\subset C$, and such that

$$f'_x(x^o,u^o)(x-x^o)\leqq f'_x(x^o,u^o)(\hat{x}-x^o)+\tfrac{1}{2} \text{ if } x\in\hat{x}+Q.$$

By Theorem 2.3, $T(Q)=\bigcup_{x\in Q} T(x)$ is an open set of Y. Define W by

$$W=\{(s_1,s_2) : s_1\in R,\ s_2\in Y,\ |s_1-f(x^o,\hat{u})-f'_x(x^o,u^o)(\hat{x}-x^o)-1|<\tfrac{1}{2},\ s_2\in T(Q)\}.$$

Then W is a neighborhood of $(f(x^o,\hat{u})+f'_x(x^o,u^o)(\hat{x}-x^o)+1),0)$, and $W\subset S'$. To see the latter, take any $(s_1,s_2)\in W$. Then $s_2\in T(Q)$, hence $s_2\in T(x-\hat{x})$ for some $x\in\hat{x}+Q$, so that $s_2\geqq g'_x(x^o,u^o)(x-\hat{x})$. But $g(x^o,\hat{u})+g'_x(x^o,u^o)(\hat{x}-x^o)\leqq 0$, hence

$$s_2\geqq g(x^o,\hat{u})-g'_x(x^o,u^o)(x-x^o).$$

Further we have that $s_1\geqq f(x^o,\hat{u})+f'_x(x^o,u^o)(\hat{x}-x^o)+\frac{1}{2}\geqq f(x^o,\hat{u})+f'_x(x^o,u^o)(x-x^o)$, by definition of Q. Since $(x,\hat{u})\in C\times V$, it follows that $(s_1,s_2)\in S'$. Hence $W\subset S'$, so that the interior of S' is nonempty.

Since (0,0) is a boundary point of the convex set S', and since int $S'\neq\emptyset$, we can now apply basic theorems of convex programming, see e.g. Ponstein[4, Theorems 3.8.14 and 3.8.11], and conclude that

$$f(x^o,u^o)=\inf_{x,u}\{f(x^o,u)+f'_x(x^o,u^o)(x-x^o)+\lambda^o y\ :\ g(x^o,u)+g'_x(x^o,u^o)(x-x^o)\leqq y,\ (x,u)\in C\times V\}.$$

This gives $\lambda^o\geqq 0$ (as otherwise the infimum would be minus infinity), and hence that

$$f(x^o,u)+f'_x(x^o,u^o)(x-x^o)+\lambda^o g(x^o,u)+\lambda^o g'_x(x^o,u^o)(x-x^o)\geqq f(x^o,u^o) \text{ if } (x,u)\in C\times V.$$

Take $(x,u)=(x^o,u^o)$, then $\lambda^o g(x^o,u^o)\geqq 0$, but also $\lambda^o g(x^o,u^o)\leqq 0$, hence $\lambda^o g(x^o,u^o)=0$. Take $u=u^o$, then

$$f'_x(x^o,u^o)(x-x^o)+\lambda^o g'_x(x^o,u^o)(x-x^o)\geqq 0 \text{ if } x\in C.$$

Take $x=x^o$, then

$$f(x^o,u)+\lambda^o g(x^o,u)\geqq f(x^o,u^o) \text{ if } u\in V.$$

End of proof.

Relaxing some of the conditions of Theorem 6.1.

1) As before, it is sufficient to assume that $\hat{x}$ is in the relative interior of C, rather than in its interior.
2) In the proof it is sufficient to know that the closure of S' is convex, rather than S' itself. Therefore, it is sufficient to assume that the closures of C and S are convex rather than C and S themselves.
3) Instead of assuming that $g'_x(x^o,u^o)X+K=Y$, we may assume that $g'_x(x^o,u^o)X+K=L$, where L is a linear subspace of finite co-dimension (so that there is a finite dimensional subspace M of Y, such that L and M span Y).
4) Finally, we may require that not C, but D, or its closure, is convex, where D is the cone of internal directions of C at x^o (see Definition 3.4). Then in the minimum principle C must be replaced by D. In the linearized version of the given problem, see Part A of the proof of Theorem 6.1, we must replace C by D as well. If (x^c,u^o) would not be an optimal solution of the linearized problem, we would again get that

$$g(x(\tau)+o(\tau),u(\tau))\leqq 0 \text{ for some } o(\tau).$$

Since $x'\in D$, it follows that $\tau(x'+W)+(1-\tau)x^o\subset C$ for all τ, $0<\tau\leqq\tau^o$ for some $\tau^o>0$, and some neighborhood W of $0 \in X$. Take τ so small that $\tau(x'+o(\tau)/\tau)+(1-\tau)x^o \in C$, so that $x(\tau)+o(\tau) \in C$, and reason as before.

Slater-type regularity conditions.

The proof of Theorem 6.1. can be greatly simplified if the nonnegative cone K of Y has a nonempty interior, and if for some $(\hat{x},\hat{u}) \in C \times V$,

$$(6.1) \qquad g(x^o,\hat{u})+g'_x(x^o,u^o)(\hat{x}-x^o)=e<0 \text{ or } e \in \text{int } (-K).$$

Theorem 6.2. *Theorem* 6.1 *remains true, if the conditions 1) to 4) are replaced by* (6.1).

Proof. We only give a sketch of the proof, for which convex processes are not necessary. Details can be found in Ponstein [5].If (x^o,u^o) would not be a solution of the linearized problem, letting $(\tilde{x},\tilde{u})$, (x',u') and $(x(\tau),u(\tau))$ be as before, we would have that

$$g(x^o,u(\tau))+g'_x(x^o,u^o)(x(\tau)-x^o)<\tau\gamma e \text{ if } \tau \text{ is small enough,}$$

so that, by the generalized mean value theorem, $g(x(\tau),u(\tau))<0$ if τ is small enough. It would follow that $(x(\tau),u(\tau))$ is a feasible solution of the original problem, so that $f(x(\tau),u(\tau)) \geqq f(x^o,u^o)$, which would contradict $f(x^o,u(\tau))+f'_x(x^o,u^o)(x(\tau)-x^o) <f(x^o,u^o)-\tau\varepsilon$. Hence $(0,0)$ is again a boundary point of S', defined in Part B of the proof of Theorem 6.1. Since $g(x^o,\hat{u})+g'_x(x^o,u^o)(\hat{x}-x^o)<0$, there is a neighborhood Q of $0 \in Y$ such that

$$-g(x^o,\hat{u})-g'_x(x^o,u^o)(\hat{x}-x^o)+Q \subset K,$$

and defining the open set W by

$$W=\{(s_1,s_2) \ : \ s_1 \in R, \ s_2 \in Y, \ |s_1-f(x^o,\hat{u})-f'_x(x^o,u^o)(\hat{x}-x^o)-1|<\tfrac{1}{2}, \ s_2 \in Q\}$$

if follows that $W \subset V$, so that int $V \neq \emptyset$. As before, this leads to the desired results. End of proof.

Not only convex processes are not necessary, but it also not necessary to invoke any open mapping theorem, generalized or not. This is because the set Q in the proof of Theorem 6.2 is a subset of Y, whereas in the proof of Theorem 6.1 it is a subset of X, which has to be transformed into a subset of Y. As in the end an open subset of Y is required, some sort of open mapping theorem is necessary, in case a Slater-type regularity condition is not applicable. We may, therefore, conclude that *necessary optimality conditions involving a Slater-type regularity condition are essentially easier to prove than such conditions but involving 'mapping-onto'-type regularity conditions, as in Theorem 6.1.*

Relaxations of some of the conditions of Theorem 6.2.

1) Again we may assume that not C and S but their closures are convex.
2) We may also assume that not C, but E (or its closure) is convex, where E is any first-order approximation of C at x^o, which is defined as follows.

Definition 6.3. E is a *first-order approximation of* C *at* x^o, if E is a cone with x^o as its apex, and if $E=\{x \ : \ \tau x+(1-\tau)x^o+o(\tau) \in C$ for some $o(\tau)$, $\tau \geqq 0\}$.

Since the cone of internal directions D is a first-order approximation of C, we see that the present relaxation is not stricter than the analagous relaxation of the assumptions of Theorem 6.1. This strengthens our conclusion that Slater-type regularity conditions are essentially weaker than 'mapping-onto'-type regularity conditions.

Remarks regarding the condition that $\lim u(\tau)=u(0)=u^o$.

In Ioffe, Tihomirov [1], results similar to Theorem 6.2 are given. Their convexity requirements are weaker than ours, but their main results in this respect are Fritz-John-type theorems, which means that there is an additional, nonnegative, multiplier for the objective function. As for any practical application it would seem that this additional multiplier should be positive, extra conditions are required to ensure this. In [1] these extra conditions imply our convexity requirements, except that it is not required that $\lim u(\tau)=u(0)=u^o$. Further, the number of scalar inequality constraints must be finite, and a condition like $x\in C$ is not taken up. Since convex processes are not used in [1], no theorems like Theorem 6.1 are presented, so we may see Theorem 6.1 as a complement to results in this book, and to a lesser extent the same is true for Theorem 6.2.

Let us now consider the condition $\lim u(\tau)=u(0)=u^o$ in more detail. If both, f and g are *separable*, that is to say if

$$f(x,u)=f_1(x)+f_2(u) \text{ and } g(x,u)=g_1(x)+g_2(u),$$

then this condition may be omitted, as can easily be verified. It follows that e.g. *linear optimal control* problems can be handled by either Theorem 6.1 or 6.2. These theorems can, however, also be applied to certain nonlinear optimal control problems. Consider, for instance, (see Section 3),

$$g(x,u)(t)=x(t)-\xi_0-\int_0^t \psi(x(\sigma),u(\sigma),\sigma)d\sigma.$$

Our convexity requirements would imply that given $(x^o,u^o)\in C\times V$, $\tilde{u}\in V$, $0\leqq\tau\leqq 1$, for some $u(\tau):t\mapsto u(\tau,t)$,

$$\int_0^t \psi(x^o(\sigma),u(\tau,\sigma),\sigma)d\sigma \leqq \tau\int_0^t \psi(x^o(\sigma),\tilde{u}(\sigma),\sigma)d\sigma+(1-\tau)\int_0^t \psi(x^o(\sigma),u^o(\sigma),\sigma)d\sigma.$$

If V is a suitable set of piece-wise continuous functions, then we can find $u(\tau)$ such that $\lim u(\tau)=u(0)=u^o$ if $\tau\downarrow 0$, and even satisfy this inequality with the equality sign. For then we may set, given any small subinterval $[a,a+\delta]$ of length δ,

$$u(\tau,\sigma)=\tilde{u}(\sigma) \text{ if } a\leqq\sigma<a+\tau\delta \text{ , and } u(\tau,\sigma)=u^o(\sigma) \text{ if } a+\tau\delta\leqq\sigma\leqq a+\delta.$$

The fact that only the closure of S, not S itself must be convex, may be helpful here.

Final remark.

It goes without saying that we may split the constraint $g(x,u)\leqq 0$ into $g_1(x,u)\leqq 0$, which may include equality constraints, and $g_2(x,u)\leqq 0$, which may not include equality constraints, and apply simultaneously 'mapping-onto'-type regularity conditions to $g_1(x,u)\leqq 0$ and Slater-type regularity conditions to $g_2(x,u)\leqq 0$.

7. MIXED MINIMIZATION AND CONCAVITY.

If (x^o,u^o) is an optimal solution of

$$\inf_{x,u}\ \{f(x,u) : g(x,u)\leqq 0,\ x\in C,\ u\in V\}$$

then, trivially, x^o is an optimal solution of

$$\inf_x\ \{f(x,u^o) : g(x,u^o)\leqq 0,\ x\in C\}$$

and we may expect that if f and g are differentiable and other suitable conditions hold, there exists some $\lambda^o\in Y^*$, $\lambda^o\geqq 0$, $\lambda^o g(x^o,u^o)=0$, such that

$$f'_x(x^o,u^o)(x-x^o)+\lambda^o g'_x(x^o,u^o)(x-x^o)\geqq 0 \text{ if } x\in C,$$

which is part of the minimum principle. Now assume that both f and g are separable, hence let

$$f(x,u)=f_1(x)+f_2(u), \text{ and } g(x,u)=g_1(x)+g_2(u).$$

Also assume that for each $u\in V$ there is an $x\in C$ such that $g(x,u)\geqq 0$ (not $\leqq 0$), and that both f and g are concave with respect to x at x^o (for each u). Then we have that

$$0\leqq g(x,u)\leqq g(x^o,u)+g'_x(x^o,u)(x-x^o) \text{ (by concavity)}$$

or

$$0\leqq g(x^o,u)+g'_x(x^o,u^o)(x-x^o) \text{ (by separability)}$$

or

$$0\leqq\lambda^o g(x^o,u)+\lambda^o g'_x(x^o,u^o)(x-x^o)\geqq\lambda^o g(x^o,u)-f'_x(x^o,u^o)(x-x^o).$$

Again by separability and concavity, this becomes

$$0\leqq\lambda^o g(x^o,u)-f'_x(x^o,u)(x-x^o)\leqq\lambda^o g(x^o,u)-f(x^o,u^o)+f(x^o,u)$$

so that

$$f(x^o,u)+\lambda^o g(x^o,u)\geqq f(x^o,u^o) \text{ if } u\in V,$$

which is the remaining part of the minimum principle. Hence we have shown our last theorem.

Theorem 7.1. *Let* X *and* Y *be Banach spaces, and let* U *be a topological vector space. Let* $f:X\times U\to R$ *and* $g:X\times U\to Y$ *be Fréchet differentiable at a point* $x^o\in X$, *and let* $C\subset X$, $V\subset U$. *Let* f *and* g *be separable, and be concave at* x^o *with respect to* x. *Assume that for each* $u\in V$ *there is an* $x\in C$ *such that* $g(x,u)\geqq 0$. *Then, if for some* $\lambda^o\geqq 0$, $\lambda^o g(x^o,u^o)=0$,

$$f'_x(x^o,u^o)(x-x^o)+\lambda^o g'_x(x^o,u^o)(x-x^o)\geqq 0 \text{ if } x\in C, \textit{ for some } u^o\in V,$$

it follows that

$$f(x^o,u)+\lambda^o g(x^o,u)\geqq f(x^o,u^o) \textit{ if } u\in V.$$

As we may consider the conclusion of this theorem as the kernel of the minimum principle, we see that this kernel can be derived from separability and concavity requirements.

REFERENCES.

[1] IOFFE, A.D., TIHOMIROV, V.M., Theory of extremal problems, North Holland, 1979.

[2] LIUSTERNIK, L., SOBOLEV, V., Elements of functional analysis, Ungar, 1961.

[3] NIEUWENHUIS, J.W., About a local approximation theorem and an inverse function theorem, J. Austral. Math. Soc., Series B, 22, 1980, 185-192.

[4] PONSTEIN, J., Approaches to the theory of optimization, Cambridge, 1980.

[5] PONSTEIN, J., On the essence of the maximum principle, University of Groningen, internal report, 93(OR-8206).

[6] PONSTEIN, J., A local approximation theorem applied to mixed programming, University of Groningen, internal report, 84-04-OR.

[7] ROBINSON, S.M., Normed convex processes, Trans. Amer. Math. Soc. 174, 1972, 127-140.

[8] ROBINSON, S.M., An inverse function theorem for a class of multivalued functions, Proc. Amer. Math. Soc. 41, 1973, 211-218.

[9] ROBINSON, S.M., Stability theory for systems of inequalities, PART II: Differentiable nonlinear systems, SIAM. J. Num. Anal. 13, 1976, 497-513.

[10] ROCKAFELLAR, R.T., Convex analysis, Princeton, 1970.

[11] TUY, H., On the convex approximation of nonlinear inequalities, Math. Operationsforsch. u. Statist. 5, 1974, 451-464.

SOME LINEAR PROGRAMS IN PROBABILITIES AND THEIR DUALS

W.K. Klein Haneveld

1. INTRODUCTION.

An important contribution of R.T. Rockafellar to the theory of optimization is his general theory of conjugate duality [27]. He provides a beautiful framework for the analysis of the optimality conditions of rather abstract mathematical programming problems in terms of dual programs. In this article we are dealing with the special case of *linear* programming duality. It has its own flavour, since the dual problem of a linear programming problem is easily formulated in an explicit form. The optimality conditions boil down to primal and dual feasibility and complementary slackness.

Many problems in mathematics can be formulated as abstract linear programming problems, and it is to be expected that dualization provides insight in the structure of the problem. In the following we consider a number of linear programs in which the decision variables are *probability measures*. These problems are well-known: they stem from robustness analysis where the probability distribution is partly unknown (e.g. moment problems) or from dynamic programming where the state probabilities are partly controllable. In each case the nonnegativity of the variables is obvious, and also the linearity of objective function and constraint mappings is quite natural, since many operations on probabilities such as expectations (but also passing to a marginal distribution, or composition with a fixed transition probability) are linear. In fact, there is a natural duality between 'measures' and 'functions' provided by 'integration', the integral being interpreted as bilinear form. Of course, unless the underlying spaces are finite, the linear programs in probabilities are infinite dimensional. However, since Rockafellar formulated his duality theory in the framework of general topological vector spaces it is not unreasonable to expect that this stumble-block can be overcome. This point of view is the basic idea of this article. Since the theory of moment problems and that of stochastic dynamic programming are well-established, new results are not to be expected. Instead, the question is: is it possible to derive the well-known results by stressing the linear structure of the problem and by applying duality theory? It will appear that although 'the' dual linear programs are easily formulated, the main difficulty is the precize choice of the spaces. Of course, this choice is crucial for the applicability of general duality theorems. As a matter of fact, whereas the dual problems have precisely the form which is to be expected, the simple application of general duality theory appears to be of restricted value, since the verification of sufficient conditions for normality and stability only is possible under too restrictive assumptions. That is, by exploiting the special properties of each of the considered problems direct proofs of duality results have been given under weaker conditions than probably is possible just as a corollary of abstract linear programming duality.

In section 2 we build up the framework of abstract linear programming duality, and in the next section some basic features of linear programs in probability measures are formulated. Section 4 deals with the generalized moment problem and its dual. Special attention is paid to the moment problem with (lattice) bounds on the probabilities. The relation with Chebychev approximation and the Neyman-Pearson lemma is indicated. The marginal problem, a generalization of the classical transportation problem, is the subject of section 5. It is related to many inequalities in probability theory. In section 6 we study the finite horizon stochastic dynamic programming problem. The last section provides the conclusions.

2. DUAL LINEAR PROGRAMS.

In the formulation of linear programming problems inequality conditions are fundamental. An inequality "$\geqq$" in a vector space X over the reals is defined by

$$x_1 \geqq x_2 \quad \text{iff} \quad x_1-x_2 \in X_+ \quad (x_1,x_2 \in X)$$

where X_+ is any convex cone in X with apex 0. The vectors of $X_+ = \{x \in X\text{:}x \geqq 0\}$ are called positive, and X_+ is called the positive cone. Extreme specifications as $X_+ := X$, in which case the condition $x \geqq 0$ is not restrictive, and $X_+ := \{0\}$, in which case 0 is the only positive vector, are not excluded. The inverse relation is of course defined by $x_1 \leqq x_2 \Leftrightarrow x_2 \geqq x_1$, and the vectors of $-X_+ = \{x \in X\text{:}x \leqq 0\}$ are called negative. If $X_+ \cap (-X_+) = \{0\}$, $X = (X,\geqq)$ is called an ordered vector space, since the relation $\geqq$ is a partial order in that case.

In general setting, a linear programming problem is defined as an optimization problem of one of the following types

$$\text{minimize}_{x\in X}\{cx\text{: } Lx \geqq b,\ x \geqq 0\}$$

$$\text{maximize}_{x\in X}\{cx\text{: } Lx \leqq b,\ x \geqq 0\}$$

where $b \in U$, X and U are real vector spaces with fixed positive cones, $L\colon X \to U$ is a linear map and $c\colon X \to \mathbb{R}$ is a linear form on X. Notice that also the case $Lx = b$ is covered (take $U_+ = \{0\}$). Similarly, if one likes to suppress the constraints $x \geqq 0$, X_+ can be defined as X. Of course, maximization can be transformed into minimization, and $\leqq$ inequalities into $\geqq$ inequalities, but we shall deal with both standard forms since it is convenient in the treatment of duality.

A pair of real vector spaces (V,X) is called a *duality* if a bilinear form $\langle .,. \rangle\colon V \times X \to \mathbb{R}$ has been singled out. Then $\langle v,. \rangle$ is a linear form on X, for each $v \in V$, and $\langle .,x \rangle$ is a linear form on V, for each $x \in X$. A shorthand notation for the duality is $\langle V,X \rangle$.

Definition 1. *Dual pair of linear programming problems.*
The pair of linear programs

LP_1 minimize$_{x\in X}\{\langle c,x\rangle: L_1x \geqq b,\ x \geqq 0\}$

LP_2 maximize$_{y\in Y}\{\langle y,b\rangle: L_2y \leqq c,\ y \geqq 0\}$

is called a *dual pair* if their data satisfy the following conditions:

a. X,U,Y,V are real vector spaces, each with a fixed positive cone.

b. $\langle V,X\rangle$ and $\langle Y,U\rangle$ are two dualities. The dualities are compatible with the positivities:

$$\text{(1)}\qquad \begin{array}{l} x \geqq 0 \text{ and } v \geqq 0 \text{ imply } \langle v,x\rangle \geqq 0, \\ u \geqq 0 \text{ and } y \geqq 0 \text{ imply } \langle y,u\rangle \geqq 0. \end{array}$$

c. L_1: X → U and L_2: Y → V are adjoint linear maps, i.e.

$$\text{(2)}\qquad \langle L_2y,x\rangle = \langle y,L_1x\rangle \text{ for all } x \in X,\ y \in Y.$$

d. b ∈ U, c ∈ V.

If (LP_1,LP_2) is a dual pair then LP_i is called a dual of LP_{3-i}, i = 1,2. By definition it is a symmetrical relation.

Before going into more details about this definition, we first like to indicate that weak duality,

$$\text{(3)}\qquad \inf LP_1 \geqq \sup LP_2,$$

can easily be shown for arbitrary dual pairs of linear programs. Remark first that (3) holds trivially if at least one of the programs is infeasible. Further, if x and y are feasible for LP_1 and LP_2, respectively, then

$$\text{(4)}\qquad \langle c,x\rangle \geqq \langle L_2y,x\rangle = \langle y,L_1x\rangle \geqq \langle y,b\rangle$$

where the inequalities follow from (1) and the equality is given by (2). From (4) (3) is derived by infimization over x and supremization over y. With respect to optimal solutions of LP_1 and LP_2, it is clear that optimality of as well x as y is guaranteed if everywhere in (4) the equality is true, that is if the *complementary slackness* conditions hold:

$$\text{(5)}\qquad \begin{array}{l} \langle c-L_2y,x\rangle = 0, \\ \langle y,L_1x-b\rangle = 0. \end{array}$$

This trivial analysis proves the following theorem.

Theorem 2. Elementary duality theorem.
Let LP_1 *and* LP_2 *be a dual pair of linear programming problems. Then the following statements are true.*

a. $\inf LP_1 \geqq \sup LP_2$.

b. $\left.\begin{array}{l} x \textit{ is feasible for } LP_1 \\ y \textit{ is feasible for } LP_2 \\ \textit{complementary slackness holds} \end{array}\right\} \Leftrightarrow \left\{\begin{array}{l} x \textit{ is optimal for } LP_1 \\ y \textit{ is optimal for } LP_2 \\ \inf LP_1 = \sup LP_2. \end{array}\right.$

Notice that for the proof theorem 2 one does not need very detailed information on the data. In particular, no topological assumptions are made. Nevertheless, optimality is guaranteed by feasibility and complementary slackness. On the other hand, complementary slackness is only necessary for optimality if there is no duality gap ($\inf LP_1 = \sup LP_2$) and if both problems have optimal solutions. Of course, the last condition can be relaxed by looking at sequences: if $x^{(k)}$ and $y^{(k)}$ are feasible for LP_1 and LP_2, respectively, for all $k = 1,2,\ldots$ then $(x^{(k)})$ and $(y^{(k)})$ are optimal sequences and $\inf LP_1 = \sup LP_2$ iff complementary slackness holds asymptotically:

$$\begin{aligned} &\lim_{k\to\infty} \langle c - L_2 y^{(k)}, x^{(k)} \rangle = 0, \\ &\lim_{k\to\infty} \langle y^{(k)}, L_1 x^{(k)} - b \rangle = 0. \end{aligned} \tag{6}$$

In spite of its elementary nature theorem 2 is important, since it shows that one might circumvent intricate topological questions if complementary slackness can be proved in a direct way. We shall come back to this point in the sequel (see e.g. theorem 10).

It is clear that definition 1 is useless unless adjoint maps exist. Of course, any linear map $L: X \to U$ has a unique algebraic adjoint map $L': U' \to X'$ defined by $(L'u')(x) := u'(Lx)$, $x \in X$, $u' \in U'$, where U' $(,X')$ is the algebraic dual space of $U(,X)$. But it will appear to be important to allow for the possibility that V and Y are not the algebraic duals of X and U. Nevertheless it is not difficult to settle the question of existence and uniqueness of the adjoints in this general framework. In order to show this we introduce the following definitions. A duality $\langle V,X\rangle$ is called *separated* if X distinguishes points in V

$$\langle v,x\rangle = 0 \quad \forall x \in X \Rightarrow v = 0 \tag{7}$$

and if V distinguishes points in X

$$\langle v,x\rangle = 0 \quad \forall v \in V \Rightarrow x = 0. \tag{8}$$

Under (7) all $v \in V$ represent different linear forms on X so that V may be identified with a subspace of X'. Similarly, $X \subset V'$ if (8) holds.

Proposition 3. *Let* $\langle V,X\rangle$ *and* $\langle Y,U\rangle$ *be separated dualities. Then any linear map* $L_1: X \to U$ *has a unique adjoint map* $L_2: Y \to V$ *provided* V *is large enough:* $V \supset L_1'(Y)$. *Similarly, any linear map* $L_2: Y \to V$ *has a unique adjoint map* $L_1: X \to U$ *provided* $U \supset L_2'(X)$.

Proof. Given L_1, define L_2 as the restriction of L_1' to Y. This is possible since U distinguishes points in Y. The range of L_2 is a subset of V by assumption, and (2) holds. The uniqueness of L_2 follows from the assumption that X distinguishes points in V. Similarly, for any linear map $L_2: Y \to V$ there exists a unique $L_1: X \to U$ such that (2) holds, if V distinguishes points in X, if $L_2'(X) \subset U$ and if Y distinguishes points in U. □

Each linear program has at least one dual in the sense of definition 1. In order to be specific, assume that LP_1 with the data X, X_+, U, U_+, L_1, b and c has been given. Then natural choices for Y, V, L_2 are U', X', L_1' together with $Y_+ :=$ $:= \{y \in Y: y(u) \geqq 0 \ \forall u \in U_+\}$, $V_+ := \{v \in V: v(x) \geqq 0 \ \forall x \in X_+\}$. This completes the definition of LP_2 and all assumptions of definition 1 hold. Often, however, the natural definition of a dual problem is not suitable in practice, since explicit representations for dual variables are missing. Therefore it is convenient to restrict attention to suitable subspaces of the algebraic duals. Usually the choice of the spaces is made for topological reasons. Starting again with LP_1, one may choose topologies on X and U such that c and L_1 are continuous, and then define LP_2 in terms of the topological dual spaces. Then Y(,V) is defined as the linear space $U^*(,X^*)$ of all continuous linear forms on U(,X). Again, if the dualities are separated and $L_1'(U^*) \subset X^*$, then one arrives at a candidate for LP_2 (and L_1' is weakly continuous automatically [28] IV.2.1). Whether this second construction of LP_2 is appropriate, depends on the possibility of a suitable choice of the topologies. In particular, one likes to have 'nice' dual spaces together with tractable topologies. In the framework of theorem 1 the spaces Y and V are choosen a priori, and in applications supposed to be 'nice'. Given the dualities $\langle V,X\rangle$ and $\langle Y,U\rangle$ only those topologies on X and U are to be considered which satisfy $X^* = V$, $U^* = Y$. Such topologies are called *compatible* with the duality. The weakest topology on X which is compatible with the duality $\langle V,X\rangle$ is the weak topology $\sigma(X,V)$, and the strongest one is the Mackey topology $\tau(X,V)$ ([28] IV.3.3).

As said before, complementary slackness is only necessary for optimality if there is no duality gap. In fact, apart from the elementary duality theorem we only need results from advanced duality theory in order to prove *normality*: inf LP_1 = = sup LP_2. In the sequel we shall restrict attention to Rockafellar's conjugate duality theory [27]. In order to make a connection between his results and the dual pair (LP_1,LP_2) we shall first investigate the *conjugate duals* of LP_1 and LP_2. As the optimal value function of LP_1 we define the function $\varphi_1 : U \to [-\infty,\infty]$,

$$\varphi_1(u) := \inf_{x \geqq 0}\{\langle c,x\rangle : L_1x \geqq b-u\}, \ u \in U.$$

The conjugate dual problem of LP_1 is then

LP_1^* $\qquad$ $\text{maximize}_{y \in Y}(\inf_{u \in U}(\varphi_1(u)+\langle y,u\rangle))$.

Similarly, the optimal value function φ_2 of LP_2 is

$$\varphi_2(v) := \sup_{y \geqq 0}\{\langle y,b\rangle : L_2y \leqq c+v\}, \ v \in V,$$

and the conjugate dual problem of LP_2 is defined as

LP_2^* $\qquad$ $\text{minimize}_{x \in X}(\sup_{v \in V}(\varphi_2(v)-\langle v,x\rangle))$.

Proposition 4. *Let* (LP_1,LP_2) *be a dual pair of linear programs. If the positive cones* V_+, Y_+ *are induced by the bilinear forms and the positive cones* X_+, U_+, *that is if*

$$(9)\quad \begin{aligned} v \geqq 0 &\Leftarrow \langle v,x\rangle \geqq 0 \quad \forall x \geqq 0,\\ y \geqq 0 &\Leftarrow \langle y,u\rangle \geqq 0 \quad \forall u \geqq 0, \end{aligned}$$

then $LP_1^* = LP_2$. *Similarly, if* X_+ *and* U_+ *are induced by the bilinear forms and* V_+ *and* Y_+, *that is if*

$$(10)\quad \begin{aligned} x \geqq 0 &\Leftarrow \langle v,x\rangle \geqq 0 \quad \forall v \geqq 0,\\ u \geqq 0 &\Leftarrow \langle y,u\rangle \geqq 0 \quad \forall y \geqq 0, \end{aligned}$$

then $LP_2^* = LP_1$.

Remark. The inverse implications in (9) and (10) are true of course because of (1).

Proof. The objective function ψ of LP_1^* can be written as

$$\psi(y) = \inf_{x,u}\{\langle c,x\rangle+\langle y,u\rangle : x \geqq 0,\ L_1x \geqq b-u\}$$

or, with $u_0 := L_1x-b+u$, as

$$\begin{aligned}\psi(y) &= \inf_{x,u_0}\{\langle c,x\rangle+\langle y,b-L_1x+u_0\rangle : x \geqq 0,\ u_0 \geqq 0\}\\ &= \langle y,b\rangle + \inf_{x\geqq 0}\langle c-L_2y,x\rangle + \inf_{u_0\geqq 0}\langle y,u_0\rangle\end{aligned}$$

because of (2). Hence

$$\psi(y) = \begin{cases}\langle y,b\rangle & \text{if } L_2y \leqq c \text{ and } y \geqq 0\\ -\infty & \text{if } L_2y \nleqq c \text{ and/or } y \ngeqq 0;\end{cases}$$

the first statement follows from (1) and the second from (9). Since by definition $\sup_{y\in\phi}\langle y,b\rangle = -\infty$ we showed that $LP_1^* = LP_2$. The proof of $LP_2^* = LP_1$ under (10) is similar. □

In the next theorem we apply a sufficient condition for *stability* (i.e. no duality gap, and the dual extremum is attained) from conjugate convex duality theory to a dual pair of linear programs.

Theorem 5. Advanced duality theorem.

Let (LP_1,LP_2) *be a dual pair of linear programs with separated dualities.*

a. If φ_1 *is bounded above, hence continuous, in the neighbourhood of* $0 \in U$ *(in a compatible topology on* U*) and if* (9) *holds, then*

$$(11)\quad -\infty \leqq \inf LP_1 = \sup LP_2 < +\infty,$$

and the supremum is attained if it is finite.

b. If φ_2 *is bounded below, hence continuous, in the neighbourhood of* $0 \in V$ *(in a compatible topology on* V*) and if* (10) *holds, then*

$$(12)\quad \infty \geqq \sup LP_2 = \inf LP_1 > -\infty,$$

and the infimum is attained if it is finite.

Proof. a. Because of proposition 4 we conclude from (9) that $LP_2 = LP_1^*$; then (11) is a direct consequence of conjugate convex duality theory ([27] Th 17a, [25] Th. 3.8.11 and 3.8.16); both references consider only separated dualities. b. Follows by reversing signs. □

Remark. The boundedness condition on the optimal value function is not necessary for normality or even stability. Weaker sufficient conditions and even necessary and sufficient conditions have been formulated (e.g. [25],[27]). However, in applications one needs conditions which are operational; that is, it must be possible to verify whether or not they are satisfied in the problem at hand. Since we were not able to do this in the examples to be given, we restrict the attention to the more practical conditions of theorem 5.

3. LINEAR PROGRAMMING PROBLEMS IN PROBABILITIES

In this section we shall describe the common parts of the linear programs to which we like to apply the duality theory. In the next sections we shall give more details.

Moment problem. The classical moment problem can be described as:

For given subset S of $\mathbb{R}$, find a probability measure on S with prescribed first m moments, which minimizes the expectation of a given criterion function on S.

Often the criterion function is the indicator of a subset $S_o \subset S$ so that the probability of S_o is minimized.

Marginal problem.

For given spaces S_i, $i = 1,\dots,m$, find a probability measure on $S := S_1 \times S_2 \times \dots \times S_m$ with prescribed marginals on each S_i, which minimizes the expectation of a given criterion function on S.

Again, the criterion function may be an indicator function.

Stochastic dynamic programming problem.

Consider a discrete-time Markovian decision process. That is, the state of the system is random and evoluates in time in a Markovian way. At each stage of time an action is taken which influences as well the costs of that stage as the transition to the state of the next stage. Decisions on the actions are based on perfect observation of the current state, and are to be taken in such a way that the total expected costs are minimal. In terms of partly controllable probabilities the problem on a finite horizon can be defined as follows.

For given state and action spaces $S_0, A_0, S_1, A_1, \dots, S_{n-1}, A_{n-1}, S_n$ find a probability measure on $S_0 \times A_0 \times S_1 \times A_1 \dots \times S_n$ with prescribed marginal on S_0 and prescribed transition probabilities from $S_{i-1} \times A_{i-1}$ to S_i, $i = 1,\dots,n$, which minimizes the expectation of the total costs.

In each problem the decision variables are probability measures. Both the objective function and the constraints are linear. Without going into specific details of each of the problems, we shall explain why such problems fit in the framework of dual linear programs in a natural way.

First of all, we shall embed the set of probability measures on a space S in a *linear* space. Let S be an arbitrary set (in applications a Borel subset of a Euclidean space will do) and $\mathcal{S}$ a σ-algebra of subsets of S which contains the singletons (e.g. the Borel subsets of $S \subset \mathbb{R}^n$). A finite signed measure on the measurable space $(S,\mathcal{S})$ is by definition any real (hence finite) countably additive function on $\mathcal{S}$ with $x(\phi) = 0$. Because of the assumption on finiteness the space $\mathcal{M}^o(S)$ of finite signed measures on $(S,\mathcal{S})$ is a vector space, which contains the probability measures on $(S,\mathcal{S})$. In fact the probabilities are a convex subset of the natural positive cone of $\mathcal{M}^o(S)$,

$$\mathcal{M}^o_+(S) := \{x \in \mathcal{M}^o(S) : x(E) \geqq 0 \quad \forall E \in \mathcal{S}\},$$

consisting of all finite measures on $(S,\mathcal{S})$. This cone defines a partial order on $\mathcal{M}^o(S)$ since every signed measure in $\mathcal{M}^o(S)$ can be decomposed as a difference of two measures in $\mathcal{M}^o_+(S)$. In fact, $\mathcal{M}^o(S)$ is a *vector lattice* under the partial order induced by the positive cone. This means, that each pair $(x_1,x_2) \in \mathcal{M}^o(S)\times\mathcal{M}^o(S)$ has a unique least (order) upperbound $\sup(x_1,x_2) \in \mathcal{M}^o(S)$ and a unique greatest (order) lowerbound $\inf(x_1,x_2) \in \mathcal{M}^o(S)$. These statements are a direct consequence of the existence of a unique Jordan decomposition of any $x \in \mathcal{M}^o(S)$: $x = x_+ - x_-$ with $x_+,x_- \in \mathcal{M}^o_+(S)$ with the property that x_- vanishes on each measurable subset of a $\hat{S} \in \mathcal{S}$ whereas x_+ vanishes on each measurable subset of its complement in S. This minimality property makes the decomposition unique, although the positive set $\hat{S}$ is not unique, generally. It is not difficult to show that $\sup(x_1,x_2) = x_1 + (x_2-x_1)_+$ and $\inf(x_1,x_2) = -\sup(-x_1,-x_2)$ $\forall x_1,x_2 \in \mathcal{M}^o(S)$. Hence, for any $x \in \mathcal{M}^o(S)$,

$$\begin{aligned} x &= x_+ - x_- \text{ with } \inf(x_+,x_-) = 0, \\ x_+ &= \sup(0,x) \in \mathcal{M}^o_+(S), \\ x_- &= \sup(0,-x) \in \mathcal{M}^o_+(S), \\ |x| &:= \sup(x,-x) = x_+ + x_- \in \mathcal{M}^o_+(S). \end{aligned} \tag{13}$$

The measures x_+, x_-, $|x|$ are called the positive part, the negative part and the total variation of the signed measure x. The lattice structure will be helpful in calculations as will be seen in the sequel.

A natural companion of $\mathcal{M}^o(S)$ is $\mathcal{F}^o(S)$, the vector space of all $\mathcal{S}$-measurable bounded real functions on S. $\mathcal{F}^o(S)$ is also a vector lattice based on the positive cone

$$\mathcal{F}^o_+(S) := \{v \in \mathcal{F}^o(S) : v(s) \geqq 0 \quad \forall s \in S\},$$

with $\sup(v_1,v_2)(s) := \max(v_1(s),v_2(s))$, $\inf(v_1,v_2)(s) := \min(v_1(s),v_2(s))$, $s \in S$, for any $v_1,v_2 \in \mathcal{F}^o(S)$. It is easy to verify that for any $v \in \mathcal{F}^o(S)$

$$(14)\quad \begin{aligned} v &= v_+ - v_- \text{ with } \inf(v_+,v_-) = 0,\\ v_+ &= \sup(0,v) \in F_+^o(S),\\ v_- &= \sup(0,-v) \in F_+^o(S),\\ |v| &= \sup(v,-v) = v_+ - v_- \in F_+^o(S);\ |v|(s) = |v(s)|,\ s \in S. \end{aligned}$$

The definitions of $M^o(S)$ and $F^o(S)$ are such that the bilinear form

$$\langle v,x\rangle := \int_S v(s)x(ds),\ v \in F^o(S),\ x \in M^o(S),$$

defines a duality $\langle F^o(S),M^o(S)\rangle$. This duality has nice properties, as will be seen later. However, its drawback is, that we had to restrict ourselves to bounded functions in the definition of $F^o(S)$, in order to be sure that $\langle v,x\rangle$ is finite for all $v \in F^o(S)$ and $x \in M^o(S)$. Sometimes this function space is too restricted. A useful generalization is given by the following definitions. Let $e : S \to \mathbb{R}$ be a fixed S-measurable function with $e(s) \geqq 1\ \forall s \in S$. Then we define $F(S)$ as the linear space of all S-measurable real functions v on S which are bounded by a multiple of e, that is

$$(15)\qquad \sup_{s\in S}\frac{|v(s)|}{e(s)} < \infty.$$

The function e is called a *bounding function*. Furthermore we define, using the same bounding function,

$$M(S) := \{x \in M^o(S) : \int_S e(s)|x|(ds) < \infty\}$$

and the positive cones

$$F_+(S) := \{v \in F(S) : v(s) \geqq 0\ \forall s \in S\},$$

$$M_+(S) := \{x \in M(S) : x(E) \geqq 0\ \forall E \in S\}.$$

Since for $v \in F(S)$, $x \in M(S)$

$$|\int_S v(s)x(ds)| \leqq \int_S |v(s)||x|(ds) \leqq M \cdot \int_S e(s)|x|(ds) < \infty$$

it is clear that

$$(16)\qquad \langle v,x\rangle := \int_S v(s)x(ds) \quad v \in F(S),\ x \in M(S)$$

defines a duality $\langle F(S),M(S)\rangle$. If $e = 1$ this duality reduces to $\langle F^o(S),M^o(S)\rangle$ but in general we have $F(S) \supset F^o(S)$ and $M(S) \subset M^o(S)$. All statements made on the positive cones $M_+^o(S)$ and $F_+^o(S)$ together with the lattice structure of the corresponding linear spaces are also true for $M_+(S)$ and $F_+(S)$, as can be easily verified.

Proposition 6. *$\langle F(S),M(S)\rangle$ is a separated duality. Furthermore, it is compatible with the positive cones $F_+(S)$ and $M_+(S)$, and these cones induce each other in the sense of (9) and (10) by the bilinear form (16). The same statements are true for $\langle F^o(S),M^o(S)\rangle$, $F_+^o(S)$ and $M_+^o(S)$.*

Proof. If $v \in F(S)$ with $v \neq 0$, then $\exists s_0 \in S$ with $v(s_0) \neq 0$. This implies $\langle v,x_0\rangle \neq 0$ for x_0 the unit mass in s_0, which is in $M(S)$ since S contains singletons and $e(s_0)$ is finite. This proves (7). In order to prove (8), let $x \in M(S)$ with $x \neq 0$. Without loss of generality we may assume that $x(\hat{S}) \neq 0$ where $\hat{S}$ is a positive set of x. Then $\langle v_0,x\rangle \neq 0$ for v_0 the indicator function of $\hat{S}$, which is in $F(S)$ since $e(s) \geqq 1$. Hence $\langle F(S),M(S)\rangle$ is separated. The statements about the positive cones have completely similar proofs. By taking e = 1 the results on $F^\rho(S)$ and $M^\rho(S)$ follow. □

Complementary slackness conditions, see (5), have an obvious interpretation in the $\langle F(S),M(S)\rangle$-framework. Let $x \in M_+(S)$, say a probability measure, and let $v \in F_+(S)$. Then it is wellknown that

(17) $$\langle v,x\rangle = 0 \Leftrightarrow x(\{s \in S : v(s) > 0\}) = 0.$$

That is, for a given probability measure x complementary slackness of x with respect to a nonnegative function v means that v must be equal to zero, actually, almost surely with respect to x. Reversely, given a nonnegative function v complementary slackness of v with respect to a probability measure x is a condition on the support of x: x is to be restricted to the subset of S where v is equal to zero.

Summarizing, the duality $\langle F(S),M(S)\rangle$ together with the positive cones $F_+(S)$ and $M_+(S)$ is a suitable framework for setting up a dual pair of linear programs, one of which having probability measures on $(S,\mathcal{S})$ as variables. In fact, the description of the probability measures x in $M(S)$ as

$$\{x \in M(S) : x \geqq 0, \langle 1,x\rangle = 1\}$$

fits precisely in the linear framework. The same is true for objective functions of the kind $\langle c,x\rangle$ with $c \in F(S)$, representing the expected value of the real function c on S with respect to x. It will appear that also the other constraints in the three problems of the beginning of this section are linear. In the moment problem they are of the type $\langle g_i,x\rangle = b_i$ $i = 1,\ldots,m$ for fixed $g_i \in F(S)$ and $b_i \in \mathbb{R}$. In the marginal problem the linear map 'passing to the marginals' occurs. Then the constraint space is a space of measures, too. In the stochastic dynamic programming model this is also the case, but the linear map is of a more general type. In the sequel of this section we shall discuss a rather general linear map and its adjoint, which covers all we need.

Let $(S,\mathcal{S})$, e_S, $\langle F(S),M(S)\rangle$, $F_+(S)$ and $M_+(S)$ be as before, and let $(T,\mathcal{T})$, e_T, $\langle F(T),M(T)\rangle$, $F_+(T)$ and $M_+(T)$ be defined similarly, starting with a measurable space $(T,\mathcal{T})$. We shall use the same notation for the bilinear forms of both dualities, except if confusion might occur; then we make the distinction $\langle v,x\rangle_S$ and $\langle y,u\rangle_T$. A nonnegative real function g on $S\times\mathcal{T}$ will be called a *transition measure* from $(S,\mathcal{S})$ to $(T,\mathcal{T})$ if it is a measure on $\mathcal{T}$ for each fixed $s \in S$ and if it is a $\mathcal{S}$-measurable function on S for each fixed $E \in \mathcal{T}$. We use the notation $g(E|s)$, $s \in S$, $E \in \mathcal{T}$. In applications, g mostly will be a *transition probability*, $g(.|s)$ being a probability measure in that case. Then g can be interpreted as the conditional probability of a random vector η taking values in T given the value s of a random vector ξ taking values in S; this explains the notation. We define $M_+(T|S)$ as the cone of all transition measures from $(S,\mathcal{S})$ to $(T,\mathcal{T})$ which satisfy

(18) $$\sup_{s\in S} \frac{1}{e_S(s)} \int_T e_T(t)g(dt|s) =: M_0(g) < \infty.$$

As a consequence

(19) $$g(E|.) \in F_+(S) \text{ for all } E \in \mathcal{T},$$

(20) $$g(.|s) \in M_+(T) \text{ for all } s \in S.$$

For each transition measure $g \in M_+(T|S)$ two linear maps will be introduced, which will appear to be adjoint:

(21) $$\begin{aligned} &G_1: M(S) \to M(T); \text{ for } x \in M(S),\ E \in \mathcal{T}: \\ &(G_1x)(E) \equiv xg(E) := \int_S g(E|s)x(ds) = \langle g(E|.), x\rangle_S, \end{aligned}$$

and

(22) $$\begin{aligned} &G_2: F(T) \to F(S); \text{ for } y \in F(T),\ s \in S: \\ &(G_2y)(s) := \int_T y(t)g(dt|s) = \langle y, g(.|s)\rangle_T. \end{aligned}$$

In the sequel we shall show that the claims in the definitions are correct. Let us first indicate their direct probabilistic interpretations. Let ξ be a random vector taking values in S, and η a random vector taking values in T, such that the conditional distribution of η given $\xi = s$ is given by the transition probability $g(.|s)$. Then $(G_2y)(s)$ represents the conditional expectation $E[y(\eta)|\xi=s]$ whereas G_1x is the probability distribution xg of η induced by the probability distribution x of ξ and the transition probability g. In fact, the joint distribution x*g of (ξ,η) is characterized by

$$(x*g)(E_1\times E_2) := \int_{E_1} g(E_2|s)x(ds),\quad E_1 \in \mathcal{S},\ E_2 \in \mathcal{T}.$$

In §6 we shall use the notations xg and x*g.

A very special but important case occurs when the transition probability is *deterministic*, that is when

$$g(E|s) = \begin{cases} 1 \text{ if } h(s) \in E \\ 0 \text{ if } h(s) \notin E \end{cases}$$

for a measurable transition function $h: S \to T$. In this case the maps G_1 and G_2 reduce to

$$(G_1x)(E) = x(\{s \in S: h(s) \in E\}) = xh^{-1}(E),$$

$$(G_2y)(t) = y(h(t)).$$

In the deterministic case it is easily shown by transformation of variables that G_1 and G_2 are adjoint. But, not surprisingly, this holds in the general case too.

Proposition 7. *Let* $g \in M_+(T|S)$. *Then* $G_1 \equiv xg$ *and* G_2 *are welldefined and linear. Moreover, they are adjoint: for each* $x \in M(S)$ *and* $y \in F(T)$ *the identity*

$$(23) \qquad \langle y, G_1x\rangle_T = \langle G_2y, x\rangle_S$$

is true.

Proof. Because of (20), G_2y is a well-defined real function on S for each $y \in F(T)$. Moreover, since $|y(t)|/e_T(t) \leqq M < \infty \ \forall t \in T$, it follows from (18) that $|(G_2y)(s)|/e_S(s) \leqq M.M_0(g)$, so that $G_2y \in F(S)$ if G_2y is measurable. The proof of measurability is standard, by extension of the subclass for which it is known to be true (By definition it is true for y being the indicator of a set in T, hence for nonnegative measurable simple functions. From the monotone convergence theorem it follows then that G_2y is measurable for all nonnegative measurable y. Since $G_2y = G_2y_+ - G_2y_-$ the proof is complete.) so that G_2 is well-defined. Obviously it is linear. In order to show that G_1 is well-defined, we first remark that for each $x \in F(S)$ G_1x is a well-defined function on T because of (19). For $x \geqq 0$, G_1x is countably additive because of the monotone convergence theorem, and since $x = x_+ - x_-$ this holds for general $x \in M(S)$, too. Therefore, G_1x is a finite signed measure on (T,T) for each $x \in M(S)$. In order to show that $G_1x \in M(T)$, actually, we first show, for $x \in M(S)$, $y \in F(T)$

$$(23') \qquad \int_T y(t)(G_1x)(dt) = \int_S\{\int_T y(t)g(dt|s)\}x(ds).$$

Because of the linearity and the decompositions $y = y_+ - y_-$, $x = x_+ - x_-$ we only have to show (23') for $y \geqq 0$ and $x \geqq 0$. The proof is standard (Take $x \geqq 0$ fixed; (23') is true by definition for measurable indicators y, hence for measurable nonnegative simple functions y. For general nonnegative measurable y, apply the monotone convergence theorem). With (23') it follows from (18) that $G_1 x \in M(T)$ for each $x \in M(S)$. This completes the proof that G_1 is well-defined. Its linearity is trivial. Moreover, by proving (23') we showed (23), actually. □

We shall conclude this section with a number of elementary special cases for deterministic transition measures defined by a measurable transition function $h: S \to T$.

Example 1. Let $(S,\mathcal{S}) \subset (T,\mathcal{T})$, e_S is the restriction of e_T to S, and define $h(s) = s$. Then $G_1 x$ denotes $x \in M(S)$ as a signed measure on $\mathcal{T}$ rather than on $\mathcal{S}$, and $G_2 y$ is the restriction of $y \in F(T)$ to S. In particular, if $(S,\mathcal{S},e_S) = (T,\mathcal{T},e_T)$ then G_1 and G_2 are the identity maps.

Example 2. Let $(S,\mathcal{S}) = (S_1 \times S_2, \mathcal{S}_1 \otimes \mathcal{S}_2)$, $e_S(s_1,s_2) = e_{S_1}(s_1) + e_{S_2}(s_2) - 1$, $(T,\mathcal{T},e_T) = (S_1,\mathcal{S}_1,e_{S_1})$ and define $h(s_1,s_2) = s_1$. Then $G_1 x$ denotes the first marginal measure of $x \in M(S)$: $(G_1 x)(E) = x(E \times S_2)$, $E \in \mathcal{S}_1$, and G_2 describes the embedding of $y \in F(S_1)$ into $F(S_1 \times S_2)$. For obvious reasons we shall denote mappings which transform a signed measure on a product space to a marginal measure as *projections*.

4. MOMENT PROBLEMS.

Since a long time extremal values for the probability of a fixed event have been studied under partial specification (or knowledge) of the probability distribution. If only the range and the numerical value of certain moments of a one-dimensional distribution are assumed to be known, the classical moment problem as described in the previous section comes up. For an overview of the history of this problem and its generalizations we refer to Karlin and Studden [15]. In this section we shall discuss the following generalized moment problem and some of its variants.

Let $(S,\mathcal{S})$ be a measurable space, let g_i, $i = 1,\dots,m+1$, be measurable (i.e. $\mathcal{S}$-measurable) real functions on S, and let b_i $i = 1,\dots,m$ be real numbers. Then we consider the problem

MO — find a probability measure on $(S,\mathcal{S})$ which minimizes the expected value of g_{m+1} under the condition that for each $i = 1,\dots,m$ the expected value of g_i is equal to b_i.

We assume that the σ-field $\mathcal{S}$ contains the singletons. This technical assumption is not very restrictive. It is true for example if $\mathcal{S}$ contains the Borel subsets of a Borel set $S \subset \mathbb{R}^n$, or, more generally, if S is a Hausdorff topological space and $\mathcal{S}$ is the σ-field generated by the closed sets.

A natural framework for the analysis of the moment problem MO is provided by the duality of linear programming. In order to formulate the corresponding dual pair of linear programs, we take the following specifications.

(i) $X := \mathcal{M}(S)$, $V := \mathcal{F}(S)$ as in section 3, with the bounding function $e_S :=$ $:= \sup_{i=0,1,\ldots,m+1} |g_i|$ with $g_0 := 1 \in \mathcal{F}(S)$. The duality $\langle \mathcal{F}(S), \mathcal{M}(S) \rangle$ and the positive cones $\mathcal{M}_+(S)$ and $\mathcal{F}_+(S)$ are defined as in section 3.

(ii) $Y := \mathbb{R}^{1+m}$, $U := \mathbb{R}^{1+m}$ with positive cones $Y_+ := Y$, $U_+ := \{0\}$ and with the duality $\langle Y,U \rangle$ given by the usual inner product.

(iii) $L_1 : X \to U$, $(L_1 x)_i := \langle g_i, x \rangle$, $i = 0,1,\ldots,m$, $x \in X$.
$L_2 : Y \to V$, $(L_2 y)(s) := \Sigma_{i=0}^m y_i g_i(s)$, $s \in S$, $y \in Y$.

(iv) $b := (b_0, b_1, \ldots, b_m) \in U$ with $b_0 := 1$; $c := g_{m+1} \in V$.

It is easily seen that all conditions of definition 1 are satisfied, so that the linear programs

$$\mathrm{MO}_1 \qquad \text{minimize}_{x \in \mathcal{M}(S)} \left\{ \int_S g_{m+1}(s) x(ds) : \int_S g_i(s) x(ds) = b_i,\ i = 0,\ldots,m,\ x \geqq 0 \right\}$$

and

$$\mathrm{MO}_2 \qquad \text{maximize}_{y \in \mathbb{R}^{1+m}} \left\{ \Sigma_{i=0}^m y_i b_i : \Sigma_{i=0}^m y_i g_i(s) \leqq g_{m+1}(s)\ \forall s \in S \right\}$$

are a dual pair. Moreover, MO is equivalent to MO_1 since $\{x \in \mathcal{M}(S) : \int_S g_0(s) x(ds) = b_0,\ x \geqq 0\}$ is precisely the set of all probabilities on the measurable subsets of S.

Proposition 8. $(\mathrm{MO}_1, \mathrm{MO}_2)$ *is a dual pair of linear programs.* MO_1 *is equivalent to* MO. *Both dualities involved are separated. The positive cones induce each other by the bilinear form, for both dualities.*

Proof. Only the last two statements have not yet been shown. For $\langle V,X \rangle$, the statements in (7), (8), (9), (10) follow as in proposition 6; for $\langle Y,U \rangle$ they follow by direct verification. □

Proposition 8 shows, that not only the elementary duality theorem 2 but also both variants of the advanced duality theorem 5 are applicable. Of course, theorem 5^a is the most attractive here, since U is finite dimensional but V is not.

Theorem 9. *Suppose that* MO_1 *is feasible, and that the feasibility is maintained under small perturbations of the right hand sides. If* $\hat{y} \in \mathbb{R}^{1+m}$ *and* $\tilde{y} \in \mathbb{R}^{1+m}$ *exist such that*

$$(24) \qquad \Sigma_{i=0}^m \hat{y}_i g_i(s) \leqq g_{m+1}(s) \leqq \Sigma_{i=0}^m \tilde{y}_i g_i(s),\ s \in S,$$

then $\inf \mathrm{MO}_1 = \max \mathrm{MO}_2$ *and this common value is finite. Moreover,* x^* *is optimal for* MO_1 *iff it is feasible and satisfies the complementary slackness condition*

$$(25) \qquad x^*(\{s \in S : \Sigma_{i=0}^m y_i^* g_i(s) < g_{m+1}(s)\}) = 0$$

where y^* *is any optimal solution of* MO_2.

Proof. The feasibility condition and the second inequality in (24) imply that the optimal value function of MO_1 is bounded above in the neighbourhood of $0 \in U$ in the usual topology of $\mathbb{R}^{1+m}$. The first inequality in (24) shows that MO_2 is feasible so that inf $MO_1 > -\infty$. All statements in the theorem follow then from theorem 2 and 5^a. □

Theorem 9 coincides with a well-known result in moment theory, see theorem XII.2.1 in [15]. In fact, many well-known inequalities in probability theory are derived in chapter XII and XIII of [15] as so-called generalized Chebyshev inequalities, and the procedure which is proposed to derive them might be interpreted as a direct application of theorem 2. That is, instead of trying to verify the conditions in theorem 9, try to solve MO_2. This involves lower approximation of g_{m+1} by a linear combination of $g_0,\dots,g_m$. Often this approximation is exact (no slack) only in a finite number of points. The complementary slackness condition (25) indicates then that one has to restrict the attention in MO_1 to probabilities with this finite support. The many examples in [15] show, that often it is possible to find optimal solutions for MO_2 and MO_1 by just trying to satisfy primal and dual feasibility and complementary slackness.

Of course, several variations of the dual pair (MO_1, MO_2) exist for which the same results hold. For example, theorem 9 holds under unchanged conditions if minimization and maximization are reversed, together with the reversing of the inequality in MO_2. The same is true if also inequalities are considered in MO_1: just change the positive cones U_+ and Y_+ and get the same results.

The problem MO_2 can be rewritten as a one-sided Chebyshev approximation. By eliminating y_0 and the constraints, it follows that MO_2 is equivalent to

$$\underset{y\in\mathbb{R}^m}{\text{maximize}} - \|-g_{m+1}+\Sigma_{i=1}^m y_i(g_i-b_ig_0)\|_+$$

where $\|g\|_+ := \sup_{s\in S} g(s)$. Similarly, if in MO_1 the constraints $x \geqq 0$, $\langle g_0,x\rangle = 1$ are replaced by $\langle g_0,|x|\rangle \leqq 1$ then for $b_i = 0$, $i = 1,\dots,m$, MO_2 can be rewritten as the symmetrical Chebyshev approximation of g_{m+1} by linear combinations of $g_1,\dots,g_m$:

$$\underset{y\in\mathbb{R}^m}{\text{maximize}} - \| g_{m+1}-\Sigma_{i=1}^m y_i g_i \|$$

where $\|g\| := \sup_{s\in S} |g(s)|$. Hence also Chebyshev approximation fits in the framework of dual linear programs. One might object that the constraint $\langle g_0,|x|\rangle \leq 1$ is not linear. However, it is equivalent to $x = x_1 - x_2$, $\langle g_0,x_1+x_2\rangle \leqq 1$, $x_1 \geqq 0$, $x_2 \geqq 0$ and $\inf(x_1,x_2) = 0$. The last condition is the only nonlinear one, but it is not difficult to show that it can be dropped without changing the problem: if x_1 and x_2 are feasible, then the same is true for $x_1 - \inf(x_1,x_2)$ and $x_2 - \inf(x_1,x_2)$, and the objective function does not change by this transformation. We shall not work out the details of the Chebyshev approximation here, since it would not add anything really new to our exposition.

In the remainder of this section we shall consider a moment problem with bounds, as introduced by Gaivoronski [4]. In [20] we studied this problem from a linear programming point of view. The corresponding linear program is

$$\text{MOB}_1 \qquad \text{minimize}_{x \in M(S)} \left\{ \langle g_{m+1}, x \rangle : \begin{array}{l} \langle g_i, x \rangle = b_i,\ i = 0,\dots,m \\ x \geqq \underline{x},\ -x \geqq -\bar{x} \end{array} \right\}$$

where the bounds $\underline{x} \in M_+(S)$, $\bar{x} \in M_+(S)$ are fixed and satisfy $\underline{x} \leqq \bar{x}$. Furthermore we shall assume

$$(26) \qquad \begin{array}{l} g_i(s) \geqq 0 \quad \forall s \in S,\ b_i \geqq 0,\ i = 1,\dots,m, \\ \langle g_i, \underline{x} \rangle < b_i < \langle g_i, \bar{x} \rangle,\ i = 0,1,\dots,m. \end{array}$$

The first condition is equivalent to g_i being bounded from below: one might add a function $\gamma_i \cdot g_i$ to g_i and replace b_i by $b_i + \gamma b_0$ without changing MOB_1. (Recall that $g_0(s) = 1$, $b_0 = 1$ as before.) The second assumption in (26) is not very restrictive if one is not interested in infeasible problems. Indeed, if $\langle g_i, \underline{x} \rangle > b_i$ then MOB_1 does not have feasible solutions since $g_i \geqq 0$ and $x \geqq \underline{x}$ imply $\langle g_i, x \rangle > b_i$. If $\langle g_i, \underline{x} \rangle = b_i$, then the constraints $\langle g_i, x \rangle = b_i$, $x \geqq \underline{x}$ and $g_i \geqq 0$ imply $x(E) = \underline{x}(E)$ for all Borel sets $E \subset \{s \in S : g_i(s) > 0\} =: E_i$, so that in that case one might reduce MOB_1 to a problem on $S \setminus E_i$ and delete the i-th constraint. Similar remarks hold for the last inequality in (26).

The dual of MOB_1 in the sense of definition 1 is MOB_2:

$$\text{MOB}_2 \qquad \begin{array}{l} \text{maximize}_{(y,\underline{y},\bar{y})} \{ \Sigma_{i=0}^m y_i b_i + \langle \underline{y}, \underline{x} \rangle - \langle \bar{y}, \bar{x} \rangle : \\ \qquad \Sigma_{i=0}^m y_i g_i(s) + \underline{y}(s) - \bar{y}(s) = g_{m+1}(s),\ s \in S, \\ \qquad y \in \mathbb{R}^{1+m},\ \underline{y} \in F_+(S),\ \bar{y} \in F_+(S) \}. \end{array}$$

Unlike MO_2, MOB_2 has infinite dimensional variables. However, it can also be reduced to a finite dimensional concave maximization problem. In order to show this, we first remark that for any feasible $(y,\underline{y},\bar{y})$ also $(y, \underline{y} - \inf(\underline{y},\bar{y}), \bar{y} - \inf(\underline{y},\bar{y}))$ is feasible. Moreover, since $\underline{x} \leqq \bar{x}$ the objective function cannot decrease by this transformation, so that the constraint $\inf(\underline{y},\bar{y}) = 0$ may be added without loss. This gives the opportunity to eliminate $\underline{y}$ and $\bar{y}$. Therefore, MOB_2 is equivalent to

$$\text{maximize}_{y \in \mathbb{R}^{1+m}} -h(y)$$

where $h(y)$ is the finite convex function

$$h(y) := -\Sigma_{i=0}^{m} y_i b_i - \langle(\Sigma_{i=0}^{m} y_i g_i - g_{m+1})_-, \underline{x}\rangle + \langle(\Sigma_{i=0}^{m} y_i g_i - g_{m+1})_+, \bar{x}\rangle$$

$$(27) \qquad = -\langle g_{m+1}, \underline{x}\rangle - \Sigma_{i=0}^{m} y_i (b_i - \langle g_i, \underline{x}\rangle) + \langle(\Sigma_{i=0}^{m} y_i g_i - g_{m+1})_+, \bar{x}-\underline{x}\rangle$$

$$= -\langle g_{m+1}, \bar{x}\rangle + \Sigma_{i=0}^{m} y_i (\langle g_i, \bar{x}\rangle - b_i) + \langle(\Sigma_{i=0}^{m} y_i g_i - g_{m+1})_-, \bar{x}-\underline{x}\rangle.$$

The reformulations are a direct consequence of $g = g_+ - g_-$ for $g \in F(S)$.

It is easily verified that proposition 8 is still true if (MO_1, MO_2) is replaced by (MOB_1, MOB_2). Therefore it is interesting to investigate the usefulness of the duality theorems 2 and 5 for finding optimal solutions. The complementary slackness conditions are

$$(28) \qquad \begin{aligned} &\langle(\Sigma_{i=0}^{m} y_i g_i - g_{m+1})_+, \bar{x}-x\rangle = 0, \\ &\langle(\Sigma_{i=0}^{m} y_i g_i - g_{m+1})_-, x-\underline{x}\rangle = 0. \end{aligned}$$

With the notations

$$\bar{Z}(y) := \{s \in S: \Sigma_{i=0}^{m} y_i g_i(s) > g_{m+1}(s)\},$$

$$Z_0(y) := \{s \in S: \Sigma_{i=0}^{m} y_i g_i(s) = g_{m+1}(s)\},$$

$$\underline{Z}(y) := \{s \in S: \Sigma_{i=0}^{m} y_i g_i(s) < g_{m+1}(s)\},$$

we get the following expressions for primal and dual feasibility together with complementary slackness:

$$(29) \qquad \begin{aligned} &\langle g_i, x\rangle = b_i, \quad i = 0,1,\dots,m, \\ &x(E) = \bar{x}(E) \quad \text{for } E \in \mathcal{S},\ E \subset \bar{Z}(y), \\ &\underline{x}(E) \leqq x(E) \leqq \bar{x}(E) \quad \text{for } E \in \mathcal{S},\ E \subset Z_0(y), \\ &\underline{x}(E) = x(E) \quad \text{for } E \in \mathcal{S},\ E \subset \underline{Z}(y). \end{aligned}$$

Let us first consider the simple case that there are *no moment constraints* (m=0). Then it is possible to solve MOB_1 and MOB_2 by means of (29) by just applying theorem 2. In order to see this, let F be the distribution function of g_{m+1} with respect to the measure $\bar{x} - \underline{x}$:

$$F(y_0) := \int_{\{g_{m+1} < y_0\}} (\bar{x}-\underline{x})(ds) = (\bar{x}-\underline{x})(\bar{Z}(y_0)), \quad y_0 \in \mathbb{R}.$$

F is nondecreasing and continuous to the left. The limits from the right are

$$F(y_0+0) = \int_{\{g_{m+1} \leqq y_0\}} (\bar{x}-\underline{x})(ds) = (\bar{x}-\underline{x})(\bar{Z}(y_0) \cup Z_0(y_0)), \quad y_0 \in \mathbb{R}.$$

F increases from $0 = F(-\infty)$ to $F(\infty) = \langle g_0, \bar{x}-\underline{x}\rangle$. Because of the assumption $\langle g_0, \underline{x}\rangle < 1 < \langle g_0, \bar{x}\rangle$, see (26), there exists at least one $y_0^* \in \mathbb{R}$ with $F(y_0^*) \leqq 1 - \langle g_0, \underline{x}\rangle \leqq F(y_0^*+0)$. Take any such y_0^*, and choose $\alpha \in [0,1]$ such that $(1-\alpha)F(y_0^*) + \alpha F(y_0^*+0) = 1 - \langle g_0, \underline{x}\rangle$. Of course, if $\bar{x}(Z_0(y_0^*)) = \underline{x}(Z_0(y_0^*))$ any choice for α is good, but if $\bar{x}(Z_0(y_0^*)) > \underline{x}(Z_0(y_0^*))$ then

$$\alpha = \frac{1-\langle g_0,\underline{x}\rangle - F(y_0^*)}{F(y_0^*+0)-F(y_0^*)}.$$

Finally we define the measure x^* as follows:

$$\begin{aligned}
&\text{for } E \in S,\ E \subset \bar{Z}(y_0^*) &&: x^*(E) := \bar{x}(E),\\
&\text{for } E \in S,\ E \subset Z_0(y_0^*) &&: x^*(E) := (1-\alpha)\underline{x}(E) + \bar{x}(E),\\
&\text{for } E \in S,\ E \subset \underline{Z}(y_0^*) &&: x^*(E) := \underline{x}(E).
\end{aligned}$$

Theorem 10. (m=0) x^* *solves* MOB_1 *and* y_0^* *solves* MOB_2.

Proof. Direct from theorem 2 and (29). The choice of α guarantees that $\langle g_0,x^*\rangle = 1$, and the choice of y_0^* implies $\alpha \in [0,1]$. □

The solution of MOB_1 with $m = 0$ has some resemblance with the famous Neyman-Pearson lemma. The reason is, that the problem of finding a test with prescribed level of significance and maximal power also can be interpreted as a linear program, the complementary slackness conditions of which have a similar form. In order to see this, let x_0, x_1 be different but fixed probability measures on $(S,\mathcal{S})$, and α a fixed number in $(0,1)$. Then the problem is: find $S_0 \in \mathcal{S}$ such that $x_0(S_0) \leqq \alpha$ and such that $x_1(S_0)$ is maximal. By allowing for randomized tests and asking for a level of significance of exactly α one arrives at the following program, with again $g_0(S) = 1$, $\forall s \in S$,

$$NP_1 \qquad \max_{\varphi\in F^o(S)} \left\{\langle\varphi,x_1\rangle : \begin{array}{l}\langle\varphi,x_0\rangle = \alpha\\ 0 \leqq \varphi(s) \leqq g_0(s),\ s \in S\end{array}\right\}.$$

The difference with moment problems is, that now the variables are *functions* instead of measures, with bounds. Using the $\langle F^o(S), M^o(S)\rangle$ duality, we get as the dual of NP_1 in the sense of definition 1

$$NP_2 \qquad \text{minimize}_{\substack{y_0\in\mathbb{R}\\ y_1,y_2\in M^o(S)}} \left\{\alpha y_0 + \langle g_0,y_1\rangle : \begin{array}{l} y_1-y_2+y_0x_0 = x_1\\ y_1 \geqq 0,\ y_2 \geqq 0\end{array}\right\}$$

and the complementary slackness conditions are $\langle\varphi,y_2\rangle = \langle g_0-\varphi,y_1\rangle = 0$. Without loss one may eliminate y_1 and y_2 from the formulation of NP_2 by

$$y_1 = (x_1-y_0x_0)_+,\quad y_2 = (x_1-y_0x_1)_-$$

so that NP_2 is equivalent to the one-dimensional convex minimization problem

$$NP_2' \qquad \text{minimize}_{y_0\in\mathbb{R}}\, k(y_0) = \alpha y_0 + \langle g_0,(x_1-y_0x_0)_+\rangle$$

and the complementary slackness conditions become then:

$$(x_1-y_0x_0)_+(\{s \in S: \varphi(s) < 1\}) = 0,$$
$$(x_1-y_0x_0)_-(\{s \in S: \varphi(s) > 0\}) = 0.$$

These results can be made more explicit by using densities. Let $\bar{x}$ be any σ-finite measure on $(S,\mathcal{S})$ such that x_0 and x_1 have densities f_0 and f_1 w.r.t. $\bar{x}$. One might take $\bar{x}:= x_0 + x_1$, or $\bar{x}:=$ the Lebesque measure in appropriate cases. Then $(x_1-y_0x_0)_+$ has the density $(f_1-y_0f_0)_+$, and $(x_1-y_0x_0)_-$ has the density $(f_1-y_0f_0)_-$. Therefore, complementary slackness can be rewritten in the wellknown form

$$\varphi(s) = 1 \text{ for } s \in S^+_{y_0} := \{s \in S: f_1(s) > y_0f_0(s)\} \quad \text{a.e.}[\bar{x}],$$

$$0 \leqq \varphi(s) \leqq 1 \text{ for } s \in S^0_{y_0} := \{s \in S: f_1(s) = y_0f_0(s)\} \quad \text{a.e.}[\bar{x}],$$

$$0 = \varphi(s) \quad \text{for } s \in S^-_{y_0} := \{s \in S: f_1(s) < y_0f_0(s)\} \quad \text{a.e.}[\bar{x}].$$

The choice of y_0 must be such that y_0 solves NP_2': $0 \in \partial k(y_0)$. This condition can easily be worked out as the wellknown condition

$$\bar{x}(S^+_{y_0}) \leqq \alpha \leqq \bar{x}(S^+_{y_0} \cup S^0_{y_0}).$$

Finally, feasibility of φ for NP_1 can be guaranteed by choosing $\gamma \in [0,1]$ such that $\bar{x}(S^+_{y_0}) + \gamma\bar{x}(S^0_{y_0}) = \alpha$. As a consequence, we showed that the Neyman-Pearson lemma may be derived as a consequence of the elementary duality theorem 2, just as was the case for the problem MOB_1 with $m = 0$.

Let us return to the generalized moment problem MOB_1 and its dual MOB_2. The presence of *moment constraints* ($m \geqq 1$) complicates the application of theorem 2 a little bit. The complications are twofold. First of all, unlike the case $m = 0$ the assumptions (26) imply in general neither the existence of an optimal solution y^* of MOB_2 nor the feasibility of MOB_1.

Secondly, if y^* has been found there does not seem to exist an easy definition of x^* on $Z_0(y^*)$ which guarantees that all the constraints $\langle g_i,x\rangle = b_i$, $i = 0,1,\dots,m$, are satisfied. Nevertheless, if in any particular situation one is able to find a feasible x^* and a y^* such that the complementary slackness conditions (29) hold, then again theorem 2 guarantees optimality. Under somewhat restrictive assumptions we have the following result.

Theorem 11. *Consider the pair* (MOB_1,MOB_2) *with* $m \geqq 1$. *If* MOB_2 *has an optimal solution* y^* *which satisfies*

$$\bar{x}(Z_0(y^*)) = \underline{x}(Z_0(y^*)) \tag{30}$$

then MOB_1 *has the unique optimal solution* x^*

$$(31)\qquad \begin{aligned} x^*(E) &= \bar{x}(E) \text{ for } E \in S,\ E \subset \bar{Z}(y^*) \cup Z_0(y^*),\\ x^*(E) &= \underline{x}(E) \text{ for } E \in S,\ E \subset \underline{Z}(y^*) \cup Z_0(y^*). \end{aligned}$$

Proof. Optimal solutions y^* for MOB_2 are characterized by $0 \in \partial h(y^*)$. Hence also $0 \in \partial_{y_i} h(y)$, $i = 0,1,\ldots,m$. These partial subgradients can be calculated easily. Using $g_i \geqq 0$ we get

$$\partial_{y_i} h(y) = -b_i + \langle g_i,\underline{x}\rangle + \int_{\bar{Z}(y)} g_i d(\bar{x}-\underline{x}) + [0,\int_{Z_0(y)} g_i d(\bar{x}-\underline{x})].$$

Because of (30) it even follows that $\frac{\partial h}{\partial y_i}(y^*)$ exists and vanishes so that for $i = 0,1,\ldots,m$ we have

$$\int_{\bar{Z}(y^*)} g_i d(\bar{x}-\underline{x}) = b_i - \langle g_i,\underline{x}\rangle = \int_{\bar{Z}(y^*)\cup Z_0(y^*)} g_i d(\bar{x}-\underline{x}).$$

Using this relation one easily verifies that indeed x^* and y^* satisfy (29), and theorem 2 implies optimality of x^* (and y^*). □

The question arises how restrictive the conditions in theorem 11 are. Assumption (26) implies, that $h(y) \to +\infty$ if $|y_i| \to \infty$, y_j fixed for $j \neq i$, so that the coordinate directions are no directions of recession of h. Under the somewhat stronger condition that h does not recede at all, h has bounded level sets ([26], theorem 27.1.d) and the existence of an optimal y^* is guaranteed. As far as condition (30) is concerned, it holds for example if a measure μ on $(S,\mathcal{S})$ exists such that $\mu(Z_0(y^*)) = 0$ and such that $\bar{x}-\underline{x}$ has a density with respect to μ (e.g. if $S \subset \mathbb{R}^n$, μ might be the Lebesgue measure).

Let us now examine how the advanced duality theorem 5 can be used to improve theorem 11. Now topology comes in. Application of theorem 5^a does not look very attractive, since there does not seem to exist a handy topology on $\mathcal{M}(S)$ which is compatible with the duality $\langle \mathcal{F}(S),\mathcal{M}(S)\rangle$. The norm topology, $\|x\| := |x|(S)$, is too strong since it allows for nonmanageable continuous linear functionals. Theorem 5^b provides a better framework. Let us assume that

(32) S is a *compact metric* space (whose σ-field $\mathcal{S}$ is generated by the closed subsets); and $g_1,g_2,\ldots,g_{m+1}$ are continuous.

Not surprisingly, we shall restrict the attention in MOB_2 to continuous functions. That is, $\mathcal{F}(S)$ is replaced by its subspace $\mathcal{F}_c(S)$ of all continuous functions. Since under (32) the bounding function $e_S := \sup(g_0,|g_1|,\ldots.|g_{m+1}|)$ is bounded and continuous itself, it may be replaced by simply g_0. Therefore, $\mathcal{F}_c(S)$ can be identified with the normed space $\mathcal{C}(S)$ of all bounded continuous functions on S, with the supremum norm. The norm dual of $C(S)$ is precisely the space of all finite signed measures $M^0(S)$ which coincides here with $M(S)$.

Theorem 12. *Assume that* (32) *holds, and replace in* (MOB_1,MOB_2) $F(S)$ *by* $F_c(S)$. *Then* MOB_1 *has optimal solutions if it has feasible solutions. Moreover, if* x^* *is any feasible solution for* MOB_1, *it is optimal iff*

$$\lim_{k\to\infty} \langle(\Sigma_{i=0}^m y_i^{(k)} g_i - g_{m+1})_+, \bar{x}-x^*\rangle = 0$$

$$\lim_{k\to\infty} \langle(\Sigma_{i=0}^m y_i^{(k)} g_i - g_{m+1})_-, x^*-\underline{x}\rangle = 0$$

where $(y^{(k)})$ *is any minimizing sequence for* h.

Proof. We shall show that theorem 5^b applies. Clearly the dualities involved are separated, and (10) holds. For the optimal value function φ_2 of MOB_2 we have

$$(33) \qquad \varphi_2(v) \geqq -\langle|g_{m+1}|,\bar{x}\rangle - \langle|v|,\bar{x}\rangle, \quad v \in C(S).$$

This follows easily by replacing g_{m+1} by $g_{m+1}+v$ in MOB_2, and setting each y_i equal to zero. Therefore, φ_2 is bounded below in the neighbourhood of 0 in the norm topology of $C(S)$, since $\langle|v|,\bar{x}\rangle \leqq \|v\|.\|\bar{x}\|$. From theorem 5^b we conclude therefore that there is no duality gap, and that MOB_1 has optimal solutions (since both problems are feasible the common optimal value is finite). Thanks to theorem 2 with (6) and (28) the proof of theorem 12 is complete.□

Under the assumptions of theorem 12 it is not guaranteed that MOB_2 has an optimal solution, or, equivalently, that the function h attains its minimum. The function h is convex and bounded below since MOB_1 is feasible, and it goes to infinity in the coordinate directions because of (26), but still it is possible that its infimum is not attained.

The compact metriziability of S is assumed in order to be certain that $M(S)$ is the normed dual of $C(S)$. If S is just a normal topological space, then two complications arise. First, continuous functions are not necessarily bounded, so that $C(S)$ does not coincide with $F_c(S)$ any more. But even in the case that it is justified to consider only bounded continuous functions, another complication arises: the norm topology on $C(S)$ is not compatible any more with the $\langle C(S),M^o(S)\rangle$ duality. If S is not compact, then $M^o(S)$ is only a subspace of the normed dual of $C(S)$: the latter contains, apart from the signed measures of $M^o(S)$ which by defintion are *countably* additive set functions, also purely finitely additive set functions ([3], theorem IV.6.2). We prefer to avoid finite additive set functions, not only because they can not easily be handled, but also because the problem MOB_1 would loose its probabilistic meaning. Therefore, we like to use a topology weaker than the norm topology, which is compatible with the $\langle C(S),M^o(S)\rangle$ duality. On the other hand, this topology should be strong enough to make $v \to <|v|,\bar{x}>$ continuous: this guarantees that φ_2 is bounded below in the neighbourhood of $0 \in C(S)$ in this topology (see (33)). Such a topology exists as we shall show. Moreover, the first complication can be dealt with simultaneously.

Theorem 13. *Theorem* 12 *still holds if the compact metriziability assumption on* S *in* (32) *is dropped.*

Proof. Define the following family of semi-norms on $F_c(S)$: $\{p_x: x \in M_+(S)\}$ with $p_x(v) := <|v|,x>$. This family generates a local convex topology T on $F_c(S)$. Since every semi-norm is then continuous, it follows from (33) that φ_2 is bounded below in the T-neighbourhood of 0. We shall now show, that T is compatible with the duality $<F_c(S), M(S)>$. First of all, recall that a linear functional T on $F_c(S)$ is T-continuous iff it is bounded by a semi-norm, that is

$$\exists x \in M_+(S) \text{ s.t. } |T(v)| \leqq <|v|,x> \quad \forall v \in F_c(S).$$

Since for any $x \in M(S)$ and $y \in F_c(S)$ we have $|<v,x>| \leqq <|v|,|x|>$, it follows that for any $x \in M(S)$ the linear functional $v \to <v,x>$ is T-continuous. Hence $M(S) \subset F_c(S)^*$. In order to prove the reverse inclusion, let T be any T-continuous linear functional on $F_c(S)$. Then an $x_0 \in M_+(S)$ exists with $|T(v)| \leqq <|v|,x_0> \; \forall v \in F_c(S)$. Since we like to change from the $<F_c(S), M(S)>$ duality to $<C(S), M^o(S)>$ we introduce the following linear maps

$$G_1: M(S) \to M^o(S): (G_1x)(E) := \int_E e_S(s)x(ds), \; E \in S, \; x \in M(S),$$

$$G_2: C(S) \to F_c(S): (G_2\hat{v})(s) := e_S(s)\hat{v}(s), \; s \in S, \; \hat{v} \in C(S).$$

Since

$$\int \hat{v}(s)(G_1x)(ds) = \int \hat{v}(s)e_S(s)x(ds), \; \hat{v} \in C(S), \; x \in M(S),$$

it follows that $<\hat{v}, G_1x> = <G_2\hat{v}, x> \quad \forall \hat{v} \in C(S) \quad \forall x \in M(S)$ so that G_1 and G_2 are adjoint. Now define the linear functional $\hat{T}$ on $C(S)$ by $\hat{T}(\hat{v}) := T(G_2\hat{v})$, $\hat{v} \in C(S)$. Then

$$|\hat{T}(\hat{v})| = |T(G_2\hat{v})| \leqq <|G_2\hat{v}|, x_0> = <G_2|\hat{v}|, x_0> = <|\hat{v}|, G_1x_0>.$$

This means, that $\hat{T}$ can be interpreted as a continuous linear functional on the subspace of all bounded continuous functions in $L^1(\hat{x}_0)$, with $\hat{x}_0 := G_1x_0 \in M_+^0(S)$. This subspace is dense, so that $\hat{T}$ can be extended to a continuous linear functional on $L^1(\hat{x}_0)$. As a consequence, it has the representation

$$\hat{T}(\hat{v}) = \int_S \hat{v}(s)\rho(s)\hat{x}_0(ds), \quad \hat{v} \in L^1(\hat{x}_0) \supset C(S),$$

with a $\rho \in L^{\infty}(\hat{x}_0)$. Define $\hat{x}_1(E) := \int_E \rho(s)\hat{x}_0(ds)$, $E \in \mathcal{S}$; then $\hat{x}_1 \in M^0(S)$ and $\hat{T}(\hat{v}) = \langle\hat{v},\hat{x}_1\rangle$. Moreover, with $x_1(E) := \int_E (e_S(s))^{-1}\hat{x}_1(ds)$ we have that

$$\int_S e_S(s)|x_1|(ds) = \int_S |\hat{x}_1|(ds) \leqq \|\rho\| . \int_S |\hat{x}_0|(ds) < \infty$$

so that $x_1 \in M(S)$ and $G_1x_1 = \hat{x}_1$. Since G_2 is invertable, we return to T via $T(v) = \hat{T}(G_2^{-1}v)$. Putting things together we find

$$T(v) = \hat{T}(G_2^{-1}v) = \langle G_2^{-1}v,\hat{x}_1\rangle = \langle G_2^{-1}v,G_1x_1\rangle$$
$$= \langle G_2G_2^{-1}v,x_1\rangle = \langle v,x_1\rangle,\ v \in F_c(S).$$

Since $x_1 \in M(S)$ the proof is complete. □

5. THE MARGINAL PROBLEM.

Let $(S_i,\mathcal{S}_i)$ be measurable spaces such that the singletons are contained in $\mathcal{S}_i$, and let p_i be probability measures on $(S_i,\mathcal{S}_i)$, $i = 1,\dots,m$. Define $S := S_1 \times S_2 \times \dots \times S_m$ and let $\mathcal{S}$ be the σ-field on S generated by $\mathcal{S}_1 \times \mathcal{S}_2 \times \dots \times \mathcal{S}_m$, and let c be a $\mathcal{S}$-measurable real function on S. These data determine the marginal problem defined in §3:

MA $$\underset{x \in M_+^o(S)}{\text{minimize}} \{\int_S c(s)x(ds) : x(S) = 1,\ \text{proj}_i x = p_i \quad i = 1,\dots,m\}.$$

Here $M_+^o(S)$ is, as before, the cone of finite measures on $(S,\mathcal{S})$ and the condition $x(S) = 1$ guarantees that only probabilities are considered. The linear map $\text{proj}_i : M^o(S) \to M^o(S_i)$ transforms each finite signed measure x on $(S,\mathcal{S})$ into its marginal on $(S_i,\mathcal{S}_i)$:

$$(\text{proj}_i x)(E) := x(B_1 \times B_2 \times \dots \times B_m) \text{ with } B_i := E_i,\ B_j := S_j,\ j \neq i;\ E_i \in \mathcal{S}_i.$$

Of course, the condition $x(S) = 1$ is redundant, since $\text{proj}_i x = p_i$ implies $x(S) = (\text{proj}_i x)(S_i) = p_i(S_i) = 1$ since each p_i is a probability. In order that MA is a well-posed problem it is necessary to assure that its objective exists for all feasible x, for example by allowing infinite values and defining $\infty - \infty = \infty$. The last definition destroys the linearity. In the sequel we shall restrict ourselves to marginal problems for which $\int c(s)x(ds)$ is finite for all feasible x, by making assumptions on c and p_i in terms of bounding functions. Let for $i = 1,\dots,m$ e_i be a bounding function on S_i, and use this bounding function in the definition of the duality $\langle F(S_i),M(S_i)\rangle$ as given in §3. Let $\langle F(S),M(S)\rangle$ be the duality determined by the bounding function e, defined by $e(s_1,\dots,s_m) := \Sigma_{i=1}^m e_i(s_i)$. If we assume that

(34) $$p_i \in M(S_i) \quad i = 1,\dots,m$$

then each x which is feasible for MA is in $M(S)$, actually, since

$$\int_S e(s)|x|(ds) = \Sigma_{i=1}^m \int_S e_i dx = \Sigma_{i=1}^m \int_{S_i} e_i d(\text{proj}_i x) = \Sigma_{i=1}^m \langle e_i, p_i\rangle < \infty$$

so that in MA one might replace $M^o(S)$ by $M(S)$. If furthermore it is assumed that

(35) $\qquad c \in F(S)$

it is clear that $\int_S c(s)x(ds) = \langle c,x\rangle$ is finite (even bounded) on the feasible set of MA.

Assumptions (34) and (35) make it possible to put MA into the linear programming duality framework of § 2. Define the following pair of linear programs

$$\text{MA}_1 \qquad \text{minimize}_{x\in M(S)}\left\{\langle c,x\rangle: \begin{array}{ll} \text{proj}_i x = p_i & i = 1,\dots,m \\ x \geqq 0 & \end{array}\right\},$$

$$\text{MA}_2 \qquad \text{maximize}_{y_i\in F(S_i)}\left\{\Sigma_{i=1}^m \langle y_i,p_i\rangle: \begin{array}{l} \Sigma_{i=1}^m y_i(s_i) \leqq c(s_1,\dots,s_m) \\ \text{for all } (s_1,\dots,s_m) \in S \end{array}\right\}.$$

Proposition 14. *Assume* (34) *and* (35). *Then* MA_1 *is equivalent to* MA. *Moreover,* $(\text{MA}_1,\text{MA}_2)$ *is a dual pair of linear programs. Both dualities involved are separated. The positive cones induce each other by the bilinear form, for both dualities.*

Proof. The equivalence of MA and MA_1 has been shown already. That $(\text{MA}_1,\text{MA}_2)$ is a dual pair, follows with the translations

(i) $\langle V,X\rangle := \langle F(S),M(S)\rangle$; $V_+ := F_+(S)$, $X_+ := M_+(S)$.

(ii) $\langle Y,U\rangle := \langle F(S_1)\times F(S_2)\times\dots\times F(S_m),\ M(S_1)\times M(S_2)\times\dots\times M(S_m)\rangle$ with
$\langle y,u\rangle := \Sigma_{i=1}^m \langle y_i,u_i\rangle$, $y_i \in F(S_i)$, $u_i \in M(S_i)$; $U_+ := \{(0,0,\dots,0)\}$, $Y_+ := Y$.

(iii) $L_1: X \to U$, $(L_1x)_i := \text{proj}_i x$, $x \in F(S)$.
$L_2: Y \to V$, $(L_2y)_i(s_1,\dots,s_m) := y_i(s_i)$, $(s_1,\dots,s_m) \in S$.

It is easily verified that indeed $L_1(X) \subset U$ since for $x \in M(S)$ $\text{proj}_i x \in M(S_i)$:

$$\int e_i d|\text{proj}_i x| \leqq \int e_i d(\text{proj}_i |x|) = \int e_i d|x| \leqq \langle e,|x|\rangle < \infty$$

The dualities are compatible with the positivities (proposition 6, and a direct proof, respectively); L_1 and L_2 are adjoint (end of § 3). Also the remaining statements follow from proposition 6, or directly. □

How restrictive are assumptions (34) and (35)? The question is: for a given measurable function c and for given probabilities p_i, do there exist bounding functions e_i such that (34) and (35) hold? In general the answer is negative. If this is the case because MA is not well-posed we do not mind. But also well-defined marginal problems exist for which the answer is negative. A typical example arises if $c \geqq 0$ and $\int c d(p_1\times p_2\times\dots\times p_m) = +\infty$. Therefore the $(\text{MA}_1,\text{MA}_2)$

framework is not suitable for all marginal problems. Nevertheless, in many situations it can be applied. Of course, if c is bounded (which occurs for example if one likes to minimize the probability of a fixed event; or if every S_i is compact and c is continuous) $e_i = 1$ will do. If c is unbounded but "not growing too fast" then non-trivial bounding functions may be relevant. A typical example occurs, if $S_i = \mathbb{R}$, if $\int |s_i|^k dp_i(s_i) < \infty$ and if $|c(s_1,\dots,s_m)| \leq M\Sigma_{i=1}^m(1+|s_i|^k)$ for an integer k and a finite M. We conclude that the (MA_1,MA_2) framework is suitable for many but not for all marginal problems. In fact, the assumptions (34) and (35) can be interpreted as regularity conditions: they are sufficient to guarantee that MA, MA_1 and MA_2 are well-defined, and that inf MA_1 and sup MA_2 are finite. Indeed, they imply that the objective functions of MA, MA_1 and MA_2 are well-defined and finite for all feasible solutions, and that MA_2 has feasible solutions. The feasibility of MA_1 is trivial: take $x_0 := p_1 \times p_2 \times \dots \times p_m$, the probability measure on S determined by the marginals under assumption of independence.

Similar as for the moment problem, the dual problem MA_2 of the moment problem MA_1 can be interpreted as an approximation problem in a function space: MA_2 is the problem of finding a *separable* lower bound y for c with maximal expected value with respect to any probability measure x which is feasible for MA_1, e.g. the independent measure x_0. In this context complementary slackness of a pair of feasible solutions (x,y) for (MA_1,MA_2) means, that the measure x must be restricted to the subset of S on which the approximation is exact, i.e.

(36) $$x(\{s = (s_1,\dots,s_m) \in S: \Sigma_{i=1}^m y_i(s_i) < c(s_1,\dots,s_m)\}) = 0.$$

This explains for example, that the independent solution x_0 can only be optimal if c itself is separable a.s.$[x_0]$. In this trivial case every feasible solution for MA_1 is optimal in fact. If the criterion function c is not separable one may expect heavy dependence between the marginals.

It is interesting to notice, that the linear program MA_1 has a special structure: it is a generalization of the well-known transportation problem. Indeed, for m = 2 and S_1 and S_2 finite sets, the problem MA_1 is reduced to the transportation problem, and its usual linear programming dual is precisely the corresponding reduction of MA_2. Some results on the problem MA_1 can be seen as generalizations of results on the transportation problem; for example concerning the optimality of the NW Rule solution (see [21]).

In different degrees of generality, marginal problems have been studied extensively in probability theory. We refer to [21], [17] for some references. More often than not, the basic approach can be described in terms of linear programming duality, although this is not always formulated explicitly, or even recognized. In the (MA_1,MA_2) framework, this approach is merely an application of the elementary duality theorem 2: try to find feasible solutions (x,y) for

(MA_1,MA_2) which satisfy complementary slackness; if one succeeds then optimality is guaranteed. This approach worked also in the analysis of robustness against dependence in projectplanning, see [21]: every result on the worst-case joint distribution can be derived by relying on the elementary duality theorem only (in stead of the duality theorem 1 in [21]); in that case $S_i \subset \mathbb{R}$ and the bounding functions $e_i(s_i) = 1+|s_i|$ do the job.

Let us now analyze the use of the advanced duality theorem 5 for the dual pair (MA_1,MA_2). Proposition 14 shows, that all conditions of as well theorem 5^a as theorem 5^b are satisfied if we are able to find a compatible topology in which the optimal value function is bounded above/below in the neighbourhood of 0. As for moment problems with bounds, application of theorem 5^a does not seem very atractive, neither with the norm topology nor with weak topologies. Moreover, it is clear that the optimal value function $\varphi_1(u) = \infty$ for all perturbations u which do not satisfy $u_1(S_1) = u_2(S_2) = \dots = u_m(S_m)$. Let us therefore consider theorem 5^b. It is not difficult to prove, that the optimal value function φ_2 is bounded below in the norm topology of $V = F(S)$, defined by

$$\|v\| := \sup_{(s_1,\dots,s_m)\in S} \frac{|v(s_1,\dots,s_m)|}{\Sigma_{i=1}^m e_i(s_i)} .$$

Indeed, since $y_i := -\|c\|e_i$, $i = 1,\dots,m$, is feasible for MA we know that $\sup MA_2 \geqq -\|c\|\ \Sigma_{i=1}^m \langle e_i,p_i\rangle$, so that the optimal value function φ_2 of MA_2 satisfies

$$\varphi_2(v) \geqq -\|c+v\|.\Sigma_{i=1}^m \langle e_i,p_i\rangle \geqq -M_1-M_2\|v\|,\ v \in V.$$

On the other hand, without additional assumptions it is not true, that the norm topology is compatible with the $\langle F(S),M(S)\rangle$ duality. Since $C(S)^*$ is isometrically isomorfic to the space of all regular countable additive set functions on the Borel sets of S if S is a compact Hausdorff space [3, p. 265] and since every countable set function is regular if S is metric [23, p. 27], [1, prop. 7.17] the following assumptions are appropriate

(37) S_i is compact metric for $i = 1,\dots,m$;
c is continuous in the product topology.

As a consequence, S is compact metric and c is bounded, so that without loss we may replace e_i by 1 $\forall_i$. Therefore (37) implies (34) and (35).

<u>Theorem 15</u>. *Assume that* (37) *holds, and replace in* (MA_1,MA_2) $F(S_i)$ by $C(S_i)$, $F(S)$ *by* $C(S)$ *and* e_i *by* 1. *Then* MA_1 *and* MA_2 *have optimal solutions, and they are com-*

pletely characterized by feasibility and complementary slackness (36).

Proof. Everything follows from applications of theorem 5^b and 2 and the discussion above, except the solvability of MA_2. But this is a direct consequence of the fact that a continuous function attains its maximum on a nonempty closed subset of a compact set. □

The application of the advanced duality theorem 5 to the marginal problem has led us to theorem 15. It is closely related to the results in [5]. We have not been able to apply theorem 5 under weaker conditions than (37), using weak topologies for example. This does not mean that theorem 15 gives the weakest conditions under which min MA_1 = max MA_2 can be proved. At the contrary: in recent papers [16], [17] Kellerer gives very weak conditions as well for normality, inf MA_1 = sup MA_2, as for solvability of each of the problems. His approach is completely different from ours. He does not prove any result for general dual linear programs, but exploits the special structure of the marginal problem and its dual extensively. It is shown that for fixed tight probability measure p_i the optimal values inf MA_1 (using the lower integral) and sup MA_2 are continuous functions of c, in several topologies. Using these continuity properties, normality is shown for semi-continuous functions c, and this is extended by means of a functional version of Choquet's capacitability theorem to normality for a wide class of functions. Some of the results are:

inf MA_1 = sup MA_2 if c has a separable integrable upperbound, and if -c is a Suslin function;

min MA_1 = sup MA_2 if c is lower semi-continuous (without any boundedness assumption);

sup MA_2 = max MA_2 if c has a separable integrable upperbound (without any measurability assumption), provided sup $MA_2 > -\infty$.

For the intricate details we refer to [16], [17]. It is not likely that similar strong duality results can be proved for general linear programming problems in probabilities.

6. THE STOCHASTIC DYNAMIC PROGRAMMING PROBLEM.

In the beginning of section 3 we gave a rough description of a discrete-time Markovian decision process. We shall now give the details of this model. Since in the linear programming duality framework the choice of the spaces is crucial, we consider rather general state and action spaces. On the other hand we restrict ourselves to a finite horizon. The infinite horizon case allows for a similar analysis with similar results.

The *data* of the Markovian decision process are the following.

Stages : $j = 0,1,\dots,n$.

State spaces : measurable spaces $(S_j,\mathcal{S}_j)$, $\mathcal{S}_j$ containing all singletons, $j = 0,\dots,n$.

Action spaces : measurable spaces $(A_j,\mathcal{A}_j)$, $\mathcal{A}_j$ containing all singletons, $j = 0,\dots,n-1$.

Constraint sets: $D_j(s_j) \in \mathcal{A}_j$, $s_j \in S_j$, $j = 0,1,\dots,n-1$, describing the admissible actions at stage j in state s_j.
$D_j := \{(s_j,a_j): a_j \in D_j(s_j)\} \in \mathcal{S}_j \otimes \mathcal{A}_j$, $j = 0,1,\dots,n-1$; its σ-algebra $\mathcal{D}_j$ of measurable subsets is the restriction of $\mathcal{S}_j \otimes \mathcal{A}_j$. It is assumed that each D_j contains the graph of a measurable mapping from S_j to A_j. For $j = n$ we define $(D_n,\mathcal{D}_n) :=$ $:= (S_n,\mathcal{S}_n)$.

Probabilities : (initial) probability measure p_0 on $(S_0,\mathcal{S}_0)$; transition probabilities (see §3 for the definition) p_j from $(D_{j-1},\mathcal{D}_{j-1})$ to $(S_j,\mathcal{S}_j)$, $j = 1,\dots,n$.

Cost functions : Borel measurable functions $c_j: D_j \to \mathbb{R}$, $j = 0,1,\dots,n$.

The fundamental issue of the optimization problem which will be formulated in terms of the data of the Markovian decision process is the concept of *policy*. By definition, a (randomized Markovian) policy δ is a sequence $(\delta_0,\delta_1,\dots,\delta_{n-1})$ of admissible transition probabilities δ_j from $(S_j,\mathcal{S}_j)$ to $(A_j,\mathcal{A}_j)$; δ_j is called admissible if $\delta_j(D_j(s_j)|s_j) = 1$ $\forall s_j \in S_j$. A policy δ is called *deterministic* or nonrandomized if each δ_j is deterministic (see §3): in that case δ_j can be identified with a measurable mapping (by a slight abuse of notation also denoted by) $\delta_j: S_j \to A_j$ with $\delta_j(s_j) \in D_j(s_j)$ $\forall s_j \in S_j$; such δ_j is called an (admissible) *decision rule* at stage j. The assumptions on the constraint sets imply that deterministic policies exist. A fortiori, the set of all policies, denoted by Δ, is nonempty.

It is well-known that an initial probability measure and suitable transition probabilities define a unique simultaneous probability on the product space. Therefore, any policy δ determines, together with the given (transition) probabilities p_j, a probability measure q_δ on $S_0 \times A_0 \times \dots \times S_{n-1} \times A_{n-1} \times S_n$, characterized by

$$q_\delta(\bar{S}_0\times\bar{A}_0\times\dots\times\bar{S}_{n-1}\times\bar{A}_{n-1}\times\bar{S}_n) = \int_{\bar{S}_0}\{\int_{\bar{A}_0}\{\dots$$

$$\{\int_{\bar{S}_{n-1}}\{\int_{\bar{A}_{n-1}} p_n(\bar{S}_n|s_{n-1},a_{n-1})\delta_{n-1}(da_{n-1}|s_{n-1})\}p_{n-1}(ds_{n-1}|s_{n-2},a_{n-2})\}\dots$$

$$\}\delta_0(da_0|s_0)\}p_0(ds_0)$$

where $\bar{S}_j(,\bar{A}_j)$ is an arbitrary set in $S_j(,A_j)$. The probability measure q_δ will sometimes be denoted by $p_0*\delta_0*\ldots*p_{n-1}*\delta_{n-1}*p_n$, and mathematical expectation with respect to q_δ is denoted by E_δ.

We are now ready to formulate the stochastic dynamic programming problem:

SDP find a policy $\delta \in \Delta$ which minimizes $c_\delta := E_\delta \Sigma_{j=0}^n c_j$.

Of course, this problem is only well-defined if the mathematical expectation of $\Sigma_{j=0}^n c_j$ exists: apart from measurability conditions which are taken care of in the specification of the data also sufficient finiteness has to be assumed in order to avoid meaningless expressions as $\infty - \infty$. For this reason it is necessary to assume

$$\text{for each } \delta \in \Delta,\ E_\delta(\Sigma_{j=0}^n c_j)_- < \infty \text{ and/or } E_\delta(\Sigma_{j=0}^n c_j)_+ < \infty.$$

Usually stronger assumptions are appropriate, such as

$$\text{(38)}\quad \begin{array}{l} \text{a. } E_\delta(c_j)_- < \infty,\ j = 0,1,\ldots,n, \text{ for all } \delta \in \Delta, \text{ and/or} \\ \text{b. } E_\delta(c_j)_+ < \infty,\ j = 0,1,\ldots,n, \text{ for all } \delta \in \Delta\ . \end{array}$$

One of the advantages is, that under (38) $\Sigma_{j=0}^n E_\delta c_j$ is a well-defined extended real number which is equal to c_δ, $\forall \delta \in \Delta$. As a consequence, c_δ can be calculated by recursion (see e.g. [1] lemma 8.1):

$$\text{(39)}\quad \begin{array}{l} v_n^\delta := c_n, \text{ and for } j = n-1,\ldots,0: \\ v_j^\delta := \int_{A_j} \{c_j(.,a) + \int_{S_{j+1}} v_{j+1}^\delta(s) p_{j+1}(ds|.,a)\} \delta_j(da|.); \\ c_\delta := \int_{S_0} v_0^\delta dp_0. \end{array}$$

Although under the condition (38) expressions as $\infty - \infty$ are avoided effectively, infinite values for integrals remain possible. Since our duality scheme does not allow for infinite-valued bilinear forms, we shall be more restrictive than (38), see (41c).

The linear programming formulation of SDP starts with the idea, to take certain marginal distributions of q_δ as decision variables rather than the policy δ itself. For any $\delta \in \Delta$ we define the marginal probability measures

$$x_j^\delta := \text{proj}_{S_j \times A_j} q_\delta,\ j = 0,1,\ldots,n-1,$$
$$x_n^\delta := \text{proj}_{S_n} q_\delta,$$

where the same projection notation is used as in §5. Since δ is admissible, x_j^δ

is a probability on D_j, actually. The probabilities x_j^δ satisfy certain linear relations:

$$\mathrm{proj}_{S_0} x_0^\delta = p_0,$$

$$\mathrm{proj}_{S_i} x_i^\delta = x_{i-1}^\delta p_i, \; i = 1,\ldots,n.$$

Here, as in (21), the composition $x_{i-1}^\delta p_i$ is defined by

$$(x_{i-1}^\delta p_i)(E) := \int_{D_{i-1}} p_i(E|s_{i-1},a_{i-1})dx_{i-1}^\delta(s_{i-1},a_{i-1}), \; E \in \mathcal{S}_i.$$

The linear relations stipulate, that $x_0^\delta, x_1^\delta, \ldots, x_n^\delta$ must agree with each other, and with the given initial and transition probabilities $p_0, p_1, \ldots, p_n$, as far as the distributions on $S_0, S_1, \ldots, S_n$ are concerned. Since c_δ can be written in terms of x_j^δ, we are led to the reformulation of SDP into a linear program in $x_0^\delta, \ldots, x_n^\delta$. In order to define suitable spaces for the x_j^δ, and in order to settle the finiteness of expected costs, we assume:

(40) for each $S_j(,A_j)$ a bounding function $e_{S_j}(,e_{A_j})$ has been choosen, and for D_j, $j < n$, we define the corresponding bounding function e_{D_j} by $e_{D_j}(s_j,a_j) := e_{S_j}(s_j) + e_{A_j}(a_j) - 1$.

Using these bounding functions, we define as in section 3 the spaces of signed measures $M(S_j)$ and $M(D_j)$, each supplied with the natural positive cones $M_+(S_j)$ and $M_+(D_j)$, and also the function spaces $F(S_j)$, $F(D_j)$ with the natural positive cones $F_+(S_j)$, $F_+(D_j)$. Again, by $M_+(S_j|D_{j-1})$ and $M_+(A_j|S_j)$ we denote the cones of transition measures which satisfy (18) with the bounding functions of (40). The following assumptions are appropriate.

(41)
a. $p_0 \in M(S_0)$, $p_j \in M_+(S_j|D_{j-1})$ $j = 1,\ldots,n$.
b. $\delta_j \in M(A_j|S_j)$, $j = 0,1,\ldots,n$.
c. $c_j \in F(D_j)$, $j = 0,1,\ldots,n$.

Of course, since p_0 and δ_j are probabilities, (41a,b) imply $p_0 \in M_+(S_0)$, $\delta_j \in M_+(A_j|S_j)$. If in (40) one chooses all bounding functions equal to 1, then (41a,b) is not restrictive at all, but (41c) forces the cost functions to be bounded. Other choices for the bounding functions may enlarge the class of feasible cost functions, but they are more restrictive for the transition probabilities, as well for the given p_j (if $e_{S_j} \neq 1$) as for the δ_j (if $e_{A_j} \neq 1$) to be determined. The next proposition shows that under (40) and (41a,b) $x_j^\delta \in M_+(D_j)$. Together with (41c) it implies that $E_\delta|c_j| = \langle|c_j|, x_j^\delta\rangle$ is finite, so that both (38a) and (38b) are true under (40) and (41), and so that $c_\delta = \Sigma_{j=0}^{n} \langle c_j, x_j^\delta \rangle$ is finite.

Proposition 16. *Assume* (40), (41a,b). *Then* $x_j^\delta \in M_+(D_j)$, $j = 0,1,\ldots,n$.

Proof. Since each x_j^δ is a (probability) measure, we only have to show that $x_j^\delta \in M(D_j)$, $j = 0,1,\ldots,n$. The proof is by induction to j. For j = 0 we have $x_0^\delta = p_0 * \delta_0$ (see §3 for the definition of *) and therefore $x_0^\delta \in M(D_0)$ since

$$\begin{aligned}\int_{D_0} e_{D_0}\, d|x_0^\delta| &= \int_{D_0} e_{D_0}\, dx_0^\delta \\ &= \int_{S_0} \{\int_{A_0} (e_{S_0}(s) + e_{A_0}(a) - 1)\,\delta_0(da|s)\} p_0(ds) \\ &\leqq (1+M)M_0 - 1\end{aligned}$$

where

$$M := \sup_{s \in S_0} (e_{S_0}(s))^{-1} \int_{A_0} e_{A_0}(a)\,\delta_0(da|s)$$

and $M_0 := \int_{S_0} e_{S_0}\, dp_0$ are finite since $\delta_0 \in M_+(A_0|S_0)$ and $p_0 \in M(S_0)$, respectively. Now assume that $x_{j-1}^\delta \in M(D_{j-1})$ for some $j \in \{1,\ldots,n\}$. Then we shall show that $x_j^\delta \in M(D_j)$. First of all, notice that $x_j^\delta = (x_{j-1}^\delta p_j) * \delta_j$. Since $p_j \in M_+(S_j|D_{j-1})$ we know that $x_{j-1}^\delta p_j \in M(S_j)$ (proposition 7). Together with $\delta_j \in M_+(A_j|S_j)$ it follows that $x_j^\delta \in M(D_j)$; similar proof as for j = 0. □

We shall now define a pair of linear programming problems, the first one of which is inspired by the discussion above.

SDP$_1$

$$\underset{x \in \Pi_{j=0}^{n} M(D_j)}{\text{minimize}} \; \Sigma_{j=0}^{n} \langle c_j, x_j \rangle$$

subject to

$$\begin{aligned} \text{proj}_{S_0} x_0 &= p_0, \\ -x_{i-1}p_i + \text{proj}_{S_i} x_i &= 0\ ,\ i = 1,\ldots,n, \\ x_j &\geqq 0\ ,\ j = 0,1,\ldots,n. \end{aligned}$$

SDP$_2$

$$\underset{y \in \Pi_{i=0}^{n} F(S_i)}{\text{maximize}} \; \langle y_0, p_0 \rangle$$

$$\begin{aligned} \text{subject to } y_j(s_j) - \langle y_{j+1}, p_{j+1}(\cdot|s_j,a_j) \rangle &\leqq c_j(s_j,a_j), \\ \forall (s_j,a_j) \in D_j,\ j &= 0,1,\ldots,n-1, \\ y_n(s_n) &\leqq c_n(s_n),\ \forall s_n \in S_n. \end{aligned}$$

Proposition 17. *Assume* (40) *and* (41a,c). *Then* (SDP_1,SDP_2) *is a dual pair of linear programs. Both dualities involved are separated. The positive cones induce each other by the bilinear form, for both dualities.*

Proof. We shall show that all conditions of definition 1 are satisfied.

a. $X := \Pi_{j=0}^{n} M(D_j)$, $V := \Pi_{j=0}^{n} F(D_j)$, both supplied with the natural positive cones $X_+ := \Pi_{j=0}^{n} M_+(D_j)$, $V_+ := \Pi_{j=0}^{n} F_+(D_j)$. $Y := \Pi_{i=0}^{n} F(S_i)$, $U := \Pi_{i=0}^{n} M(S_i)$, with positive cones $Y_+ := Y$, $U_+ := \{0\}$.

b. $\langle V,X\rangle := \Sigma_{j=0}^{n} \langle F(D_j), M(D_j)\rangle$, $\langle Y,U\rangle := \Sigma_{i=0}^{n} \langle F(S_i), M(S_i)\rangle$ are indeed dualities which are compatible with the positivities (proposition 6).

d. $\tilde{p}_0 := (p_0,0,0,\ldots,0) \in U$ and $c := (c_0,\ldots,c_n) \in V$ because of (41a,c).

c. Define $L_1: X \to U$ and $L_2: Y \to V$ by

$$(L_1x)_i := \Sigma_{j=0}^{n} (L_1)_{ij} x_j, \quad (L_2y)_j := \Sigma_{i=0}^{n} (L_2)_{ji} y_i$$

with $x = (x_0,\ldots,x_n)$, $x_j \in M(D_j)$, and $y = (y_0,\ldots,y_n)$, $y_i \in F(S_i)$, and where the maps $(L_1)_{ij}: M(D_j) \to M(S_i)$, $(L_2)_{ji}: F(S_i) \to F(D_j)$ are defined as follows:

$$\text{if } i-j = 0 : (L_1)_{ii} x_i := \text{proj}_{S_i} x_i,$$
$$((L_2)_{jj} y_j)(s_j,a_j) := y_j(s_j),$$

$$\text{if } i-j = 1 : (L_1)_{i,i-1} x_{i-1} := -x_{i-1} p_i,$$
$$((L_2)_{j,j+1} y_{j+1})(s_j,a_j) := -\langle y_{j+1}, p_{j+1}(\cdot|s_j,a_j)\rangle,$$

$$\text{if } i-j \neq 0,1: (L_1)_{ij} := 0, \ (L_2)_{ji} := 0.$$

Of course, the definitions of L_1 and L_2 are such that the constraints in SDP_1 and SDP_2 are equivalent to $L_1x \geqq \tilde{p}_0$ and $L_2y \leqq c$, respectively. We have to verify that $L_1(X) \subset U$ and $L_2(Y) \subset V$, that L_1 and L_2 are linear, and that they are adjoint. These verifications can be done for each pair $((L_1)_{ij},(L_2)_{ji})$ separately. If $i-j \neq 0,1$ these statements are trivial. If $i-j = 1$, they are a direct consequence of proposition 7, since $p_i \in M_+(S_i|D_{i-1})$ for $i \geqq 1$. Finally, if $i-j = 0$ they also follow from proposition 7 (see example 2 after it).

Finally, because of proposition 6 $\langle V,X\rangle$ is a separated duality, and V_+ and X_+ induce each other by the bilinear form. For the same reason $\langle Y,V\rangle$ is separated. Y_+ and V_+ induce each other by the bilinear form (direct proof). □

Proposition 17 guarantees that not only the elementary duality theorem 2 but also the advanced duality theorem 5 is applicable to the dual pair of linear programming problems (SDP_1,SDP_2). Before considering such applications, we shall first analyze the relation between the original stochastic dynamic programming problem

SDP and (SDP_1,SDP_2).

The relation between SDP and SDP_1 is complicated for two reasons. First, while the construction of x_j^δ from $\delta \in \Delta$ is direct, we now are faced with the reverse problem: given any probability measure x_j on $D_j \subset S_j \times A_j$, we must decompose it into a probability on S_j and a transition probability from S_j to A_j. This decomposition is possible if the spaces S_j and A_j are *Borel spaces* ([1] prop. 7.27). For our exposition it is sufficient to identify Borel spaces as complete separable metric spaces (also called Polish spaces); for a slightly more general definition and for an exposition of the properties of Borel spaces we refer to [1] and [12]. Hence we introduce the assumption:

(42) S_j and A_j are Borel spaces, and $\mathcal{S}_j$ and $\mathcal{A}_j$ are the σ-fields of the Borel sets, for all j.

Under (42) also $(D_j,\mathcal{D}_j)$ is a Borel space for all j. The second complication is assumption (41b) which has no counterpart in SDP. We introduce therefore the *restricted* stochastic dynamic programming problem

SDP' find $\delta \in \Delta' := \{\delta \in \Delta\colon \text{(41b) holds}\}$ which minimizes $c_\delta := E_\delta \Sigma_{j=0}^n c_j$.

Proposition 18. *Assume* (38), (40), (41a,c), (42). *The linear program* SDP_1 *is closely related to* SDP *and* SDP', *in the following sense:*

a. For each $\delta \in \Delta'$ *there is an* x *feasible for* SDP_1 *such that* $\langle c,x\rangle = c_\delta < \infty$.

b. For each x *feasible for* SDP_1 *there is a* $\delta \in \Delta$ *such that* $c_\delta = \langle c,x\rangle$.

c. The optimal values satisfy

$$\inf \text{SDP} \leqq \inf \text{SDP}_1 \leqq \inf \text{SDP}'. \tag{43}$$

Proof. a. Take $x := x^\delta$ as defined earlier. Then proposition 16 and the discussion before it shows what has to be proven.

b. Let $(x_0,x_1,\ldots,x_n)$ be feasible for SDP_1. Then each x_j is a probability distribution, actually, since it is positive and has total mass 1. The last statement follows by induction to j, using constraints of SDP_1 and the fact that all p_j are probabilities. Because of (42) a transition probability δ_j from S_j to A_j exists for each $j = 0,1,\ldots,n-1$ such that $x_j = (\text{proj}_{S_j} x_j)\delta_j$. Hence, $\delta := (\delta_0,\ldots,\delta_{n-1}) \in \Delta$. A direct verification shows that $x_j = x_j^\delta$, $j = 0,1,\ldots,n$.

c. This is a direct consequence of a and b. □

One is tempted to extend the proof of part b of proposition 18 by showing that the constructed δ_j are contained in $M_+(A_j|S_j)$. This would imply that $\delta \in \Delta'$, actually, so that the last inequality in (43) could be improved to the equality. Our attempts failed, however. It is possible to prove this equality, provided $\Delta' \neq \phi$, under continuity assumptions on c_j and p_j (the latter with respect to weak topology on $M(D_{j-1})$ induced by the continuous functions on D_{j-1}) by approximating δ by $\delta' \in \Delta'$, but we shall not work out the details.

Unfortunately, it is not to be expected that in general the equality inf SDP = inf SDP' holds. For example, it is possible that $\Delta' = \phi$ whereas $\Delta \neq \phi$ by assumption. But there is a simple condition which guarantees $\Delta' = \Delta$ implying inf SDP = inf SDP'.

Proposition 19. *Assume* (40), (41a,c). *Then* $\Delta' = \Delta$ *if and only if a finite* M *exists with*

(44) $$\sup_{a_j \in D_j(s_j)} e_{A_j}(a_j) \leqq M \cdot e_{S_j}(s_j), \quad \forall s_j \in S_j, \quad j = 0,1,\ldots,n-1.$$

Proof. (If). Because of (44) we have for each $\delta = (\delta_0,\ldots,\delta_{n-1}) \in \Delta$ that

$$\sup_{s_j \in S_j} (e_{S_j}(s_j))^{-1} \int_{A_j} e_{A_j}(a_j)\delta_j(da_j \mid s_j) \leqq M$$

so that $\delta_j \in M_+(A_j \mid S_j)$, $j = 0,1,\ldots,n-1$. Hence $\delta \in \Delta'$.
(Only if). Suppose that (44) is not true. Then for a certain j a sequence $((s_j^{(k)}, a_j^{(k)}),\ k = 1,2,\ldots) \subset D_j$ exists with

$$\lim_{k\to\infty} e_{A_j}(a_j^{(k)})/e_{S_j}(s_j^{(k)}) = +\infty.$$

Define a deterministic decision rule $\hat{\delta}_j$ by $\hat{\delta}_j(s_j^{(k)}) := a_j^{(k)}$, $k = 1,2,\ldots$ and arbitrary elsewhere. Then $\hat{\delta}_j \notin M(A_j \mid S_j)$, so that $\Delta' \neq \Delta$. □

Proposition 20. *Assume* (40), (41a,c), (42) *and* (44). *Then both inequalities in* (43) *are equalities, so that* SDP *and* SDP_1 *are equivalent, and* inf SDP = inf $SDP_1 < +\infty$.

Proof. A direct consequence of propositions 18 and 19. Notice that (38) is satisfied in this case, since not only (40) and (41a,c) are true but also (41b) for all $\delta \in \Delta = \Delta'$. □

Although sufficient, $\Delta' = \Delta$ is not necessary to have equalities in (43). In [18] we provide a production-inventory control model for which $\Delta' \neq \Delta$ but inf SDP' = inf SDP. On the other hand $\Delta' = \Delta$, that is (44), is true if each A_j is finite (or compact, with e_{A_j} continuous) or if $e_{A_j} \equiv 1$. In particular, by taking $e_{S_j} \equiv 1$ as well, we get

Corollary 21. *Assume* (42). *If each* c_j *is bounded, then the conclusions of proposition* 20 *hold.*

Let us now describe the relation between the stochastic dynamic programming problem SDP and the second linear program, SDP_2. To start with, notice that the constraints in SDP_1 and SDP_2 have a special feature: those in SDP_1 are 'lower-staircase' while those in SDP_2 are 'upper-staircase'. The former property reflects the fact, that the state-action probabilities x_j^δ for any $\delta \in \Delta$ in a natural way arise by *forward* recursion:

$$x_0^\delta = p_0 * \delta_0;\ \text{for } i = 1,\dots,n-1 : x_i^\delta := (x_{i-1}^\delta p_i) * \delta_i;\ x_n^\delta := x_{n-1}^\delta p_n.$$

Similarly, the latter property asks for *backward* recursion of 'cost functions'. This reveals the intimate relationship between SDP_2 and the *dynamic programming algorithm*. This algorithm, together with the associated optimality conditions, is the main tool of analysis (probably even the raison d'être) of dynamic programming theory. It defines by backward recursion the sequence of so-called *cost-to-go functions* $f_j : S_j \to [-\infty,\infty]$, $j = n,n-1,\dots,0$ by

DPA

$$f_n(s) := c_n(s),\ s \in S_n,$$
and for $j = n-1,\dots,0$
$$f_j(s) := \inf_{a \in D_j(s)} \{c_j(s,a) + \int_{S_{j+1}} f_{j+1}(t)p_{j+1}(dt|s,a)\},\ s \in S_j,$$
and finally: $\inf \text{DPA} := \int_{S_0} f_0(s)p_0(ds)$.

In these definitions the integral needs additional explanation, since f_{j+1} is not necessarily measurable and sufficiently bounded. However, under (42) it is universally measurable ([1] Corollary 8.2.1, [11] theorem 14.4), so that the integral has a unique meaning, if one also adopts the convention $-\infty+\infty = +\infty-\infty = \infty$. (For the properties of integrals under this convention we refer to [1] lemma 7.11; for our treatise the most important property is that the extended definition of an integral reduces to the classical one if the latter exists, under conditions as (38), that is.)

The importance of the algorithm is the fact that it provides optimality conditions for SDP.

<u>Proposition 22</u>. *Assume* (38). *Then*

(45) $\qquad \inf \text{DPA} \leqq \inf \text{SDP}.$

For any $\delta \in \Delta$, *define* q_δ *and* v_j^δ *as in the beginning of this section. Then*

$$\left.\begin{array}{l} v_j^\delta = f_j \quad \text{a.s. } [q_\delta] \\ j = 0,1,\dots,n \end{array}\right\} \Leftrightarrow \left\{\begin{array}{l} \delta \text{ is optimal for SDP} \\ \inf \text{DPA} = \inf \text{SDP} \end{array}\right.$$

<u>Proof</u>. Let $\delta \in \Delta$. By induction based on (39) and DPA it follows that $v_j^\delta \geqq f_j$ $j = n,n-1,\dots,0$. Therefore, $c_\delta = \int v_0^\delta dp_0 \geqq \int f_0 dp_0 = \inf \text{DPA}$. Hence $\inf_{\delta\in\Delta} c_\delta = \inf \text{SDP} \geqq \inf \text{DPA}$. If $v_0^\delta = f_0$ a.s. $[q_\delta]$ then $c_\delta = \inf \text{SDP} = \inf \text{DPA}$. Reversely, if $c_\delta = \inf \text{SDP} = \inf \text{DPA}$, then an inspection of the inequalities in the first part of the proof shows that $v_j^\delta = f_j$ a.s. $[q_\delta]$ for all j. □

The conditions $v_j^\delta = f_j$ a.s. $[q_\delta]$, $j = 0,1,\ldots,n$, are equivalent to

(46) $$\delta_j(D_j^*(s_j)|s_j) = 1,\ j = 0,1,\ldots,n-1,\ \text{a.s. } [q_\delta],$$

where, for $j = 0,1,\ldots,n-1$ and $s \in S_j$

$$D_j^*(s) := \operatorname{argmin}_{a \in D_j(s)} \{c_j(s,a) + \int_{S_{j+1}} f_{j+1}(t) p_{j+1}(dt|s,a)\}.$$

Suppose for a moment that each $D_j^*(s_j)$ is nonempty, and that each D_j^* admits a measurable selection. Then there exists a $\delta^* \in \Delta$, even a deterministic one, such that $\delta_j^*(D_j^*(s_j)|s_j) = 1$ for all j and s_j. We say such a policy is *generated by DPA*. Obviously, every policy δ^* generated by DPA is optimal. In fact, it can be shown to be 'super' optimal: it is optimal uniformly for all initial distributions p_0, and moreover, the restriction of δ^* to any tail problem is optimal for that tail problem, again uniformly for all initial distributions. Unfortunately, it is not true in general that DPA does generate a policy.

Notice that proposition 22 implies that (46) is only necessary for $\delta \in \Delta$ to be optimal for SDP if

$$\inf \text{DPA} = \inf \text{SDP}$$

is true. We shall show that this equality is closely related to the normality of the dual pair of linear programs (SDP_1, SDP_2). Let us first relate SDP_2 with DPA.

<u>Proposition 23</u>. *Assume* (38), (40), (41a,c). *Then*

(47) $$\sup \text{SDP}_2 \leqq \inf \text{DPA}.$$

Moreover, if the cost-to-go functions f_j *satisfy* $f_j \in F(S_j)$, $j = 0,\ldots,n$, *then*

$$\max \text{SDP}_2 = \inf \text{DPA}.$$

If (44) *holds*, $f_j \in F(S_j)$ *iff* f_j *is* S_j*-measurable*, $j = 0,1,\ldots,n$.

<u>Proof</u>. Without loss of generality we may assume that a feasible solution $y = (y_0,y_1,\ldots,y_n)$ for SDP_2 exists. By backward recursion, using the constraints of SDP_2, one shows that $y_i(s_j) \leqq f_j(s_j)$ for all $s_j \in S_j$, $j = n,n-1,\ldots,0$. Consequently, $\langle y_0,p_0\rangle \leqq \int f_0 dp_0 = \inf \text{DPA}$. Since y is arbitrary, $\sup \text{SDP}_2 \leqq \inf \text{DPA}$. Clearly equality holds if $f_j \in F(S_j)$ for all j, because $(f_0,f_1,\ldots,f_n)$ is feasible for SDP_2 then.

Finally we have to show that under (44)

$$\gamma_j := \sup_{s \in S_j} (e_{S_j}(s))^{-1} |f_j(s)|,\ j = 0,1,\ldots,n,$$

are finite numbers. This can be done by recursive estimation based on DPA. Let M be the number defined in (44), and define

$$M_1 := \max_j \sup_{(s,a)\in D_j} (e_{D_j}(s,a))^{-1} \langle e_{S_{j+1}}, p_{j+1}(.|s,a)\rangle,$$
$$M_2 := \max_j \sup_{d\in D_j} (e_{D_j}(d))^{-1} |c_j(d)|.$$

M_1 and M_2 are finite positive numbers because of (41a,c). From the definition of f_n in DPA we conclude that $\gamma_n \leqq M_2$. Since $|\inf E| \leq \sup |E|$ for any $E \subset [-\infty,\infty]$ we conclude from the definition of f_j, $j < n$, that

$$\begin{aligned}\gamma_j &\leqq \sup_{(s,a)\in D_j} (e_{S_j}(s))^{-1}(|c_j(s,a)| + \int |f_{j+1}(t)| p_{j+1}(dt|s,a))\\ &\leqq \sup_{(s,a)\in D_j} (e_{S_j}(s))^{-1}(M_2+\gamma_{j+1}M_1)e_{D_j}(s,a)\\ &\leqq (M_2+\gamma_{j+1}M_1)(1+M).\end{aligned}$$

Backward induction shows that each $\gamma_j < \infty$. Since $\gamma_j \geqq 0$ too, the proof is complete. □

The last result in proposition 23 indicates, that if sufficient finiteness is assumed, the equivalence between SDP_2 and DPA is 'only' a matter of measurability. Since f_j is universally measurable under (42), one can make the equivalence complete by taking for $\mathcal{S}_j$ the σ-algebra of all uniformly measurable functions on the Borel space S_j. We come back to this idea later (see proposition 30).

The propositions 18 and 20 show the relation between SDP_1 and SDP, whereas proposition 22 establishes the relations between SDP_2 and DPA. Moreover, the combination of (43), (45) and (47) reveals, that under suitable conditions

(48) $$\sup SDP_2 \leqq \inf DPA \leqq \inf SDP \leqq \inf SDP_1.$$

Therefore, the dynamic programming result inf DPA = inf SDP can be derived by showing that (SDP_1, SDP_2) has no duality gap. Then the *optimality conditions* of proposition 22 can be viewed as *consequences of complementary slackness*. In order to see this, suppose that $\sup SDP_2 = \inf SDP_1$, and let (x^*, y^*) be feasible for (SDP_1, SDP_2). The complementary slackness conditions reduce to (see (17)):

(49) $$\begin{aligned}&x_n^*(\{s_n \in S_n : y_n^*(s_n) = c_n(s_n)\}) = 1, \text{ and for } j = 0,1,\ldots,n-1:\\ &x_j^*(\{s_j,a_j) \in D_j : y_j^*(s_j) = c_j(s_j,a_j) + \langle y_{j+1}^*, p_{j+1}(.|s_j,a_j)\rangle\}) = 1.\end{aligned}$$

Because of (42) there exists a $\delta^* \in \Delta$ such that $x_j^* = x_j^{\delta^*}$, $j = 0,1,\ldots,n$ (see the proof of proposition 18b). Hence, the probabilities x_j^* are the marginals of $q_{\delta^*} = p_0*\delta_0^**\ldots*\delta_{n-1}^**p_n$. Therefore, (49) is equivalent to $y_j^* = f_j$, $j = 0,1,\ldots,n$, a.s. $[q_{\delta^*}]$, and this implies that δ^* satisfies (46).

The previous paragraph reveals the complete analogy between proposition 22 and theorem 2. Both provide sufficient optimality conditions, which are necessary, too, if there is no gap. Both have a trivial proof. In both cases, it is much more difficult to prove that there is no gap under not too restrictive assumptions. Relation (48) shows, that the dynamic programming problem is gap-free if the pair of linear programs is gap-free. Therefore, it seems only natural to derive results for the stochastic dynamic programming problem by applying advanced duality theory to the pair of linear programs. Moreover, proposition 17 shows, that all conditions of as well theorem 5^a as theorem 5^b are satisfied, if we are able to find a compatible topology in which the optimal value function is bounded above/below in the neighbourhood of 0. Just as in sections 4 and 5, application of theorem 5^a does not seem very attractive, neither with the norm topology nor with weak topologies. Moreover, it is clear that 'small' perturbations (e.g. $u_0 \in \mathcal{M}(S_0)$ with $u_0(E) < 0$ for an $E \in \mathcal{S}_0$ with $p_0(E) = 0$) may give $\varphi_1(u) = +\infty$. Let us therefore consider theorem 5^b, and analyze the optimal value function φ_2. We shall show that φ_2 is Lipschitz continuous, hence bounded below in the neighbourhood of 0, in a norm topology to be defined.

Under the assumptions (40), (41a,c) the optimal value function φ_2 of SDP_2 is well-defined, and it can be written as

$$\varphi_2(v) = \sup\{\langle y,\tilde{p}_0\rangle : y \in W(c+v)\},\ v \in V,$$

where

$$W(v) := \{y \in Y: L_2 y \leqq v\},\ v \in V.$$

Here the same definitions for Y, V, V_+, L_2, c and $\tilde{p}_0$ are used as in the proof of proposition 17. Properties of φ_2 can be derived from the study of $W(v)$. Some elementary properties of $W(v)$ are

$$W(\lambda v) = \lambda W(v);\ \lambda > 0,\ v \in V,$$
$$W(v_1) + W(v_2) \subset W(v_1+v_2);\ v_1 \in V,\ v_2 \in V,$$
$$v_1 \leqq v_2 \Rightarrow W(v_1) \subset W(v_2);\ v_1 \in V,\ v_2 \in V.$$

In the following lemma we exhibit properties of $W(v)$ which are consequences of special properties of L_2. We use the following norms on $\mathcal{F}(D_j)$, $j = 0,1,\ldots,n$ and $V = \Pi_{j=0}^n \mathcal{F}(D_j)$:

$$\|v_j\| := \sup_{(s_j,a_j)\in D_j} (e_{D_j}(s_j,a_j))^{-1} |v_j(s_j,a_j)|, \; j = 0,\dots,n-1,$$

$$\|v_n\| := \sup_{s_n \in S_n} (e_{S_n}(s_n))^{-1} |v_n(s_n)|,$$

$$\|v\| := \max_{j=0,1,\dots,n} \|v_j\| \text{ for } v = (v_0,v_1,\dots,v_n) \in V.$$

Similarly a norm on $Y = \Pi_{i=0}^{n} F(S_i)$ can be defined, but we do not need it. Finally, let $v^e \in V$, $y^e \in Y$, be given by

$$v^e := (e_{D_0}, e_{D_1}, \dots, e_{D_n}), \; y^e := (e_{S_0}, e_{S_1}, \dots, e_{S_n}).$$

Obviously, $v^e \in V_+$, $\|v^e\| = 1$. In fact v^e is the lattice supremum of all $v \in V$ with $\|v\| \leqq 1$. A similar statement characterizes y^e, at least if instead of Y_+ (= Y) the natural positive cone is used, consisting of vectors of nonnegative functions. This natural positive cone will be denoted by $\tilde{Y}_+$ and will be used in the following lemma.

Lemma 24. *a. If* $y^{(k)} \in W(v)$, $k = 1,2$, *then* $z := \sup(y^{(1)}, y^{(2)}) \in W(v)$, *where* sup *denotes the lattice supremum with respect to* $\tilde{Y}_+$.
b. Assume that (44) *holds. Then there exist positive numbers* $\eta_0,\eta_1,\dots,\eta_n$ *such that* $y^\eta := (\eta_0 y_0^e,\dots,\eta_n y_n^e)$ *satisfies*

$$-y^\eta \in W(-v^e) \text{ and } W(v^e) \subset y^\eta - \tilde{Y}_+.$$

Furthermore, $-\|v\|.y^\eta \in W(v) \subset \|v\|.y^\eta - \tilde{Y}_+$.

Proof. a. Of course, $z \in Y$ since $|z| \leqq |y^{(1)}|+|y^{(2)}|$. In order to show that $L_2 z \leqq v$, notice that $y_n^{(k)} \leqq v_n$ implies $z_n \leqq v_n$. Analogously, for $j < n$, $(s,a) \in D_j$ and $k = 1,2$

$$y_j^{(k)}(s) \leqq v_j(s,a) + \langle y_{j+1}^{(k)}, p_{j+1}(.|s,a)\rangle$$
$$\leqq v_j(s,a) + \langle z_{j+1}, p_{j+1}(.|s,a)\rangle.$$

The first inequality follows from $L_2 y^{(k)} \leqq v$, and the second from $p_{j+1} \geqq 0$. Maximization over k gives the desired result.
b. Because of (41a) and (18)

$$M_1 := \max_j \sup_{(s,a)\in D_j} (e_{D_j}(s,a))^{-1} \langle e_{S_{j+1}}, p_{j+1}(.|s,a)\rangle$$

is finite, and let M be given in (44). Let $\eta_n := 1$, and for $j = n-1,n-2,\dots,0$ define

$\eta_j := (1+\eta_{j+1}M_1)(1+M)$.
Then for each $j \in \{0,1,\ldots,n-1\}$

$$\begin{aligned} e_{D_j} + \langle y^\eta_{j+1}, p_{j+1}(.|.)\rangle &= e_{D_j} + \eta_{j+1}\langle e_{S_{j+1}}, p_{j+1}(.|.)\rangle \\ &\leqq (1+\eta_{j+1}M_1)e_{D_j} \\ &\leqq (1+\eta_{j+1}M_1)(1+M)e_{S_j} = \eta_j e_{S_j} = y^\eta_j. \end{aligned}$$

Together with $y^\eta_n = \eta_n e_{S_n} = e_{S_n} = e_{D_n}$ this implies that $L_2(-y^\eta) \leqq -v^e$. Since $-y^\eta \in Y$ we proved $-y^\eta \in W(-v^e)$.

Suppose $y \in W(v^e)$. Then $y_n \leqq e_{D_n} = e_{S_n}$, and $y_j \leqq e_{D_j} + \langle y_{j+1}, p_{j+1}(.|.)\rangle$. Backward induction shows $y_j \leqq \eta_j e_{S_j} = y^\eta_j$, $j = n,n-1,\ldots,0$, which had to be proved.

Finally, since $-y^\eta \in W(-v^e)$ we have for $v \neq 0$ that $-\|v\|.y^\eta \in W(-\|v\|.v^e)$, and because of $-\|v\|.v^e \leqq v$ we conclude $-\|v\|.y^\eta \in W(v)$; clearly, this result holds too if $v = 0$. Moreover, since $W(v^e) \subset y^\eta - \tilde{Y}_+$ and $v \leqq \|v\|v^e$ it follows that $W(v) \subset \|v\|.y^\eta - \tilde{Y}_+$ for $v \neq 0$; backward recursion shows that this inclusion is still true for $v = 0$. □

<u>Proposition 25</u>. *Assume* (40), (41a,c), (44). *Then in* SDP$_2$ *the supremum is attained. Moreover, the optimal value function* φ_2 *of* SDP$_2$ *is Lipschitz continuous:*

(50) $$|\varphi_2(v_1)-\varphi_2(v_2)| \leqq K.\|v_1-v_2\| \quad \forall v_1,v_2 \in V.$$

<u>Proof</u>. Recall that $\varphi_2(v) = \sup\{\langle y,\tilde{p}_0\rangle: y \in W(c+v)\}$. Fix $v \in V$. Lemma 24^b shows, that $W(c+v) \neq \phi$ so $\varphi_2(v) > -\infty$, and also, because of $\tilde{p}_0 \geqq 0$, that $\varphi_2(v) \leqq \leqq \|c+v\|.\langle y^\eta,\tilde{p}_0\rangle \leqq \|c+v\|.\eta_0.\langle e_{S_0},p_0\rangle < \infty$. Hence $\varphi_2(v)$ is finite. Because of that, there exists a sequence $(y^{(k)}, k = 1,2,\ldots) \subset W(c+v)$ with $\varphi_2(v) - 1/k \leqq \leqq \langle y^{(k)},\tilde{p}_0\rangle \leqq \varphi_2(v)$. Because of lemma 24^a we may assume $y^{(1)} \leqq y^{(2)} \leqq y^{(3)} \leqq \ldots$ (inequalities with respect to $\tilde{Y}_+$) without loss of generality: if necessary, replace $y^{(k)}$ by $\sup_{1\leqq h\leqq k} y^{(h)}$; this does not disturb the assumption on $\langle y^{(k)},\tilde{p}_0\rangle$ since $\tilde{p}_0 \geqq 0$. On the other hand, we know from lemma 24^b that $y^{(k)} \leqq \|c+v\|y^\eta$. Hence, $\bar{y}_j(s_j) := \lim_{k\to\infty} y^{(k)}_j(s_j)$ is finite for all j and s_j. Each $\bar{y}_j$ is measurable, so that $\bar{y} := (\bar{y}_0,\bar{y}_1,\ldots,\bar{y}_n) \in Y$ since $y^{(1)} \leqq \bar{y} \leqq \|c+v\|y^\eta$ with $y^{(1)},y^\eta \in Y$. The last condition enables us also to conclude by the dominated convergence theorem that $L\bar{y} \leqq c+v$ and that $\langle\bar{y},\tilde{p}_0\rangle = \lim_k \langle y^{(k)},\tilde{p}_0\rangle$. So we did show the existence of a $\bar{y} \in W(c+v)$ with $\langle\bar{y},\tilde{p}_0\rangle = \varphi_2(v)$.

Let us now prove the Lipschitz continuity of φ_2. Fix $v^1,v^2 \in V$ and set $\varepsilon := \|v^1-v^2\|$. Since the supremum in the definition of $\varphi_2(v^1)$ is attained, there exists a $y^1 \in Y$ with $\varphi_2(v^1) = \langle y^1,\tilde{p}_0\rangle$ and $L_2y^1 \leqq v^1$. Lemma 24^b shows that

$L_2(-y^\eta) \leqq -v^e$ so that $L_2(y^1-\varepsilon y^\eta) \leqq v^1-\varepsilon v^e \leqq v^2$; the last inequality follows from the definition of ε. Consequently $y^1-\varepsilon y^\eta \in W(v^2)$ and therefore $\varphi_2(v^2) \geqq$ $\geqq \langle y^1-\varepsilon y^\eta,\tilde{p}_0\rangle = \varphi_2(v^1)-\varepsilon\langle y^\eta,\tilde{p}_0\rangle$. By exchanging v^1 and v^2 we find also

$\varphi_2(v^1) \geqq \varphi_2(v^2) - \varepsilon\langle y^\eta,\tilde{p}_0\rangle$. Hence (50) is true with $K := \langle y^\eta,\tilde{p}_0\rangle = \eta_0\cdot\langle e_{s_0},p_0\rangle < \infty$. □

Summarizing the results we have

Theorem 26. *a. Assume* (40), (41a,c), (42) *and* (44). *Then*

(51) $$-\infty < \max SDP_2 \leqq \inf DPA \leqq \inf SDP = \inf SDP_1 < \infty.$$

b. If a topology on $V = \Pi_{j=0}^n F(D_j)$ *can be defined satisfying*

(52)
(i) $\|v\|$ *is continuous,*
(ii) *the topological dual space* V^* *of* V *is* $X := \Pi_{j=0}^n M(D_j)$,

then

(53) $$-\infty < \max SDP_2 = \inf DPA = \min SDP = \min SDP_1 < +\infty.$$

Moreover, the optimal solutions of SDP_1 *and* SDP_2 *are completely characterized by feasibility and complementary slackness. As a consequence, the conditions* (46), *inspired by* DPA, *are necessary and sufficient for optimality in the* SDP *problem.*

Proof. a. (51) follows from the propositions 20, 22, 23 and 25.
b. Using (52) and proposition 25 we derive from theorem 5^b that $\sup SDP_2 = \min SDP_1$. Proposition 18^b shows that $\min SDP = \min SDP_1$ if $\inf SDP_1 = \min SDP_1$. The remaining statements in (53) follow from (51). The last part of theorem 26 is a consequence of theorem 2, proposition 22 and the discussion at (49). □

Theorem 26 looks nice but its weak point for applications is, of course, that the existence of a topology on V which satisfies (52) has to be established. As in section 5 we are urged to come up with compactness and continuity assumptions.

(54)
S_j and A_j are compact Borel spaces, $j = 0,1,\ldots,n-1,(n)$,
D_j is closed, $j = 0,1,\ldots,n-1$,
c_j is continuous, $j = 0,1,\ldots,n$.

Since (54) implies that each D_j is compact, each c_j is bounded so that without loss we may replace each bounding function by 1.

Theorem 27. *Assume that* (54) *holds, and replace in the definition of* (SDP_1,SDP_2) $F(D_i)$ *by* $C(D_i)$, $i = 0,1,\ldots,n$, *and each bounding function by 1. Then* (40), (41a,c), (42), (44) *and* (52) *are true so that all results of theorem 26 apply.*

Proof. Direct. Use $C(D_i)^* = M(D_i)$. □

We derived theorem 27 in the framework of dual linear programs; theorem 5^b is used to prove the stability of SDP_2. Since this stability implies that SDP_1 has optimal solutions, it is not surprising that we end up with compactness and continuity assumptions. Nevertheless, the assumptions (54) are not necessary for inf DPA = min SDP: in [19, section 4] we discussed a rather general inventory control model with euclidean state and action spaces, with D_j closed but not compact, with c_j lower semicontinuous and locally bounded below. Together with some assumptions which imply that only compact subsets of $D_j(s_j)$ have to be considered and which imply certain continuity properties of the transition probabilities p_j, these assumptions are sufficient to prove inf DPA = min SDP. The proof is based on a careful analysis of DPA.

In an attempt to get rid of the compactness assumption, let us analyze what can be proved if (54) is replaced by (40), (41a,c), (42), (44) and

$$e_{D_j} \in F_c(D_j),\ c_j \in F_c(D_j),\ j = 0,1,\ldots,n,$$

where $F_c(D_j)$ is the subspace of $F(D_j)$ consisting of functions which are continuous with respect to the topology in D_j induced by the metric structure. If in SDP_2 the constraint space $V = \Pi_{j=0}^n F(D_j)$ is replaced by $V_c := \Pi_{j=0}^n F_c(D_j)$, then still its optimal value function is Lipschitz continuous with respect to the norm in V_c (inherited from the norm in V). Hence, sup SDP_2 = min $\overline{SDP_1}$ with $\overline{SDP_1}$ is the conjugate dual problem of SDP_2. $\overline{SDP_1}$ is a linear program (see [18] for an explicit description). It resembles SDP_1. The basic difference is, that its space of variables $\Pi_{j=0}^n F_c(D_j)^*$ is larger than $X = \Pi_{j=0}^n M(D_j)$, since $F_c(D_j)^*$ contains all additive (rather than countably additive) set functions on $(D_j,\mathcal{D}_j)$ which satisfy certain boundedness conditions. Since integrals with respect to bounded additive set functions do not have all nice properties which are available in the countably additive case, the constraints in $\overline{SDP_1}$ are different from those in SDP_1, but nevertheless: on X there is a complete agreement between $\overline{SDP_1}$ and SDP_1. Therefore, a possibility to show sup SDP_2 = min SDP_1 would be to decompose the optimal solution of $\overline{SDP_1}$ into its countably additive part and its purely singular part, and to find conditions which guarantee that the countably additive part solves SDP_1. Ponstein [24] has studied such conditions. However, his results refer to the decomposition of $L_\infty^*(S,\Sigma,\mu)$ rather than $L_\infty^*(S,\Sigma)$ which would be relevant for our problem. By the way, it is not to be expected that in general min $\overline{SDP_1}$ = = min SDP_1: Heilmann [9] provides an example (for an infinite horizon stationary dynamic programming problem) with min $\overline{SDP_1}$ = inf SDP_1 where the infimum is not attained.

Up to now we have been preoccupied with proving stability: sup SDP_2 = = min SDP_1. How about proving normality: sup SDP_2 = inf SDP_1? In the literature on duality theory, we did not find sufficient conditions for normality which result into easily verifiable conditions on the Markovian decision process. On the other hand, it is known from dynamic programming theory, that 'almost always' inf DPA = inf SDP holds. Typically, the proof has the following structure. In DPA, define for each $\varepsilon > 0$, each $j = n-1,\dots,0$, each $s \in S_j$ the set

$$D_j^{\varepsilon}(s) := \{a \in D_j(s) : f_j(s)+\varepsilon \geqq c_j(s,a)+\int_{S_{j+1}} f_{j+1}(t)p_{j+1}(dt|s,a)\}$$

where $f_j(s)+\varepsilon$ is replaced by $-\frac{1}{\varepsilon}$ if $f_j(s) = -\infty$. Then $D_j^{\varepsilon}(s_j) \neq \phi$. Use a measurable selection theorem in order to prove the existence of a deterministic policy δ with $\delta_j(s_j) \in D_j^{\varepsilon}(s_j)$ $\forall j\ \forall s_j$. Then it follows by recursion that if each $f_j > -\infty$

$$\text{inf SDP} \leqq \int v_0^{\delta} dp_0 \leqq \int f_0 dp_0 + n\varepsilon = \text{inf DPA}+n\varepsilon$$

so that inf SDP $\leqq$ inf DPA. In this way, using the Jankov-von Neumann analytically measurable selection theorem ([1], prop. 7.49 and 7.50), Bertsekas and Shreve prove the following result, essentially.

Proposition 28. *Assume* (38), (42). *If the set of admissible policies is enlarged with all sequences of uniformly measurable transition probabilities, then*

$$\text{inf DPA} = \text{inf SDP}.$$

Proof. Proposition 8.2 in [1]. In [1] the model is stationary, with terminal cost function $c_n = 0$. But by extending the number of stages with 1 one gets rid of the terminal cost function; and by state augmentation one reduces nonstationary models to stationary ones [1,ch.10]. The restriction to Markovian policies is justified in [1] proposition 8.11. □

Whereas we are not able to derive the dynamic programming result of proposition 28 as a corollary of a duality result, it is interesting to notice that the reverse approach works.

Proposition 29. *Assume* (40), (41a,c), (42), (44), *and replace in the definition of* SDP_2 *each* $\mathcal{S}_j$ *by the* σ*-algebra of universally measurable subsets of* S_j. *Then*

$$\max SDP_2 = \inf SDP_1.$$

Proof. Combine (the discussion after) proposition 23 with theorem 27 and proposition 28. □

In the remainder of this section we point out some relations to the existing literature which were not mentioned previously. For an extensive discussion on stochastic dynamic programming in general state and action spaces we refer to [11], [29], [30], the monograph [1] of Bertsekas and Shreve, and the references given there. The relevance of linear programming duality for dynamic programming has been recognized for a long time. We refer to [7] for an annotated bibliography. Mostly stationary models over an infinite horizon and under various objectives are considered. Complete results are known if state and action spaces are finite, e.g. [2], [22]. Yamada [32] gives results for the average cost criterion using amongst others compact subsets of $\mathbb{R}^n$ as state and action spaces. For general state and action space in a stationary infinite horizon model Heilmann [6], [8], [9] formulates linear programming models and their duals, which are comparible with SDP_2 and $\overline{SDP_1}$. Hordijk [13] gives an exposition on nonstationary finite horizon models with finite action and state spaces. Not surprisingly, in this case every dynamic programming result can be derived easily from the finitely dimensional linear programming duality theory and the simplex method. Heilmann [10] considers the general case. Our model resembles that of Heilmann [10]. However, we consider a finite horizon. More importantly, we avoid to work with noncountably additive set functions, by choosing the dual pair a priori. Moreover, our framework allows for unbounded cost functions, thanks to the use of bounding functions. Whereas to our best knowledge bounding functions are not used in the literature on the linear programming approach to dynamic programming, they are not unknown to dynamic programmers. Just as in our approach, bounding functions have been introduced as a means to guarantee finiteness of the cost-to-go functions (and their convergence in the stationary infinite horizon case [31]). Moreover, extrapolation and approximation bounds might be improved by the use of bounding functions ([12],[14]). In all these papers the dynamic programming recursion is used as the basic tool, and therefore a bounding function on the *state space* is used to get a suitable 'similarity transformation' of the state space. Typical assumptions are (in our notation)

(55) $$\sup_{(s,a)\in D_j} (e_{S_j}(s))^{-1} |c_j(s,a)| < \infty,$$

(56) $$\sup_{(s,a)\in D_j} (e_{S_j}(s))^{-1} \int_{S_{j+1}} e_{S_{j+1}}(t) p_{j+1}(dt|s,a) < \infty.$$

The linear programming approach led us to the introduction of bounding functions on the *action spaces*, too. It is interesting to notice, that (55) and (56) are equivalent to our assumptions (40), (41), (44) in the special case $e_{A_j} \equiv 1$ for all j. Therefore, our assumptions are more general, and it seems that they allow for the same results.

7. CONCLUSIONS.

We analyzed several optimization problems in probabilities from a linear programming point of view. The similarity between the various problems is striking. Under weak finiteness conditions each of them corresponds to a linear program, and for this linear program a dual linear program can be formulated which is an approximation problem in a function space. Optimality of feasible solutions for both problems is achieved if complementary slackness holds. This application of the elementary duality theorem is so trivial, that many authors apply it without realizing that it has to do with linear programming duality; they often do not formulate the dual problem and do not realize that they are looking for an optimal dual solution, too. In order to show that (asymptotical) complementarity slackness is necessary for optimality, one needs verifiable sufficient conditions for normality. For that reason we analyzed the possible application of a stability criterion provided by Rockafellar. In most cases it appeared to be profitable to consider the original problem in probabilities as the conjugate dual of the problem in function spaces. The optimal value function of the latter problem can be shown to be continuous in the norm topology, and this topology is compatible with the duality if the underlying spaces are compact metric, and the given functions are continuous. More general situations can be covered, too, if one is able to find a compatible weak topology, in which the optimal value function is bounded below. In a simple case we had success with this direct approach based on advanced duality theory. In general, however, it seems to be necessary to use the specific properties of the problem at hand in order to get the most general results. For example, one may use less strict finiteness conditions, and choose σ-algebra's which are adjusted to the problem at hand. Sometimes one needs advanced topological measure theory, and sometimes the topological considerations do not seem to be dominant in the derivation of the most general results. Anyhow, we feel that looking at optimization problems in probabilities as linear programming problems and studying their duals too, provides useful insight. This is not on strained terms with the natural fact, that using specific techniques for specific problems provides results which reach beyond the results of the general abstract duality theory.

REFERENCES.

[1] BERTSEKAS, D.P., SHREVE, S.E., Stochastic optimal control: the discrete time case (Academic Press, New York, 1978).

[2] DERMAN, C., Finite state Markovian decision processes (Academic Press, New York, 1970).

[3] DUNFORD, N., SCHWARTZ, J.T., Lineair operators part I (Interscience, New York, 1957).

[4] ERMOLIEV, Y., GAIVORONSKI, A., Duality relations and numerical methods for optimization problems on the space of probability measures with constraints on probability measures, Working paper WP-84-46, IIASA (Laxenburg 1984).

[5] GAFFKE, N., RÜSCHENDORF, L., On a class of extremal problems in statistics, Mathematische Operationsforschung und Statistik, Series Optimization

12(1981)123-135.

[6] HEILMANN, W.-R., Stochastische dynamische Optimierung als Spezialfall linearer Optimierung in halbgeordneten Vectorräumen, Manuscripta Mathematica 23(1977)57-66.

[7] HEILMANN, W.-R., Solving stochastic dynamic programming problems by linear programming - an annotated bibliography, Zeitschrift für Operations Research 22(1978)43-53.

[8] HEILMANN, W.-R., Generalized linear programming in Markovian decision processes, Bonner Mathematische Schriften 98(1978)33-39.

[9] HEILMANN, W.-R., Solving a general discounted dynamic program by linear programming, Zeitschrift für Wahrscheinlichkeitstheorie und verwandte Gebiete 48(1979)339-346.

[10] HEILMANN, W.-R., A linear programming approach to general nonstationary dynamic programming problems, Mathematische Operationsforschung und Statistik, Series Optimization 10(1979)325-333.

[11] HINDERER, K., Foundations of nonstationary dynamic programming with discrete time parameter, Lecture notes in operations research and mathematical systems 33 (Springer Verlag, Berlin, 1970).

[12] HINDERER, K., HÜBNER, G., On approximate and exact solutions for finite stage dynamic programs, in: Tijms, H.C. and Wessels, J., eds., Proceedings of the advanced seminar on Markov decision theory, Mathematical Centre Tracts 93 (Mathematisch Centrum, Amsterdam, 1977) pp. 57-76.

[13] HORDIJK, A., From linear to dynamic programming via shortest paths, in: P.C. van Baayen, D. van Dulst, J. Oosterhoff, eds., Proceedings of the Bicentennial congress of the Wiskundig Genootschap, part II, Mathematical Centre Tracts 101 (Mathematisch Centrum, Amsterdam, 1979) pp. 213-231.

[14] HÜBNER, G., Bounds and good policies in stationary finite stage Markovian decision problems, Advances in applied probability 12(1980)154-173.

[15] KARLIN, S., STUDDEN, W.J., Tchebycheff systems: with applications in analysis and statistics (Interscience, New York, 1966).

[16] KELLERER, H.G., Duality theorems for marginal problems and stochastic applications, in: Proceedings of the 7th Braşov conference on probability theory, to be published (1985).

[17] KELLERER, H.G., Duality theorems for marginal problems, to be published.

[18] KLEIN HANEVELD, W.K., The linear programming approach to finite horizon stochastic dynamic programming, Report 7903-OR, Econometrics Institute University of Groningen (Groningen, 1979).

[19] KLEIN HANEVELD, W.K., On the behavior of the optimal value operator of dynamic programming, Mathematics of operations research 5(1980)308-320.

[20] KLEIN HANEVELD, W.K., Abstract LP duality and bounds on variables, Report 84-13-OR, Econometrics Institute, University of Groningen (Groningen, 1984).

[21] KLEIN HANEVELD, W.K., Robustness against dependence in PERT: an application of duality and distributions with known marginals, Mathematical Programming Study, forthcoming (1985).

[22] MINE, H., OSAKI, S., Markovian decision processes (American Elsevier, New York, 1970).

[23] PARTHASARATHY, K.R., Probability measures on metric spaces (Academic Press, New York, 1967).

[24] PONSTEIN, J., On the use of purely finitely additive multipliers in mathematical programming, Journal of optimization theory and applications 33(1981)37-55.

[25] PONSTEIN, J., Approaches to the theory of optimization (Cambridge University Press, Cambridge, 1980).

[26] ROCKAFELLAR, R.T., Convex Analysis (Princeton University Press, Princeton NJ, 1970).

[27] ROCKAFELLAR, R.T., Conjugate duality and optimization, SIAM monograph (SIAM, Philadelphia, Pa, 1974).

[28] SCHAEFER, H.H., Topological vector spaces (Springer Verlag, New York, 1966).

[29] SHREVE, S.E., BERTSEKAS, D.P., Alternative theoretical frameworks for finite horizon discrete time stochastic optimal control, SIAM Journal on control and optimization 16(1978)953-978.

[30] SHREVE, S.E., BERTSEKAS, D.P., Universally measurable policies in dynamic programming, Mathematics of Operations Research 4(1979)15-30.

[31] WESSELS, J., Markov programming by successive approximations with respect to weighted supremum norm, Journal of mathematical analysis and applications 58(1977)326-335.

[32] YAMADA, K., Duality theorem in Markovian decision problems, Journal of mathematical analysis and applications 50(1975)579-595.

LIST OF CONTRIBUTORS

J.-B. Hiriart-Urruty
U.E.R. Mathématiques, Informatique, Gestion
Université Paul Sabatier
118, route de Narbonne
31062 Toulouse Cedex
France

W.K. Klein Haneveld
Rijksuniversiteit Groningen
Postbus 800
9700 AV Groningen
The Netherlands

J. Ponstein
Rijksuniversiteit Groningen
Postbus 800
9700 AV Groningen
The Netherlands

R. Tyrrell Rockafellar
Department of Mathematics
University of Washington
Seattle, WA 98195
USA

Vol. 157: Optimization and Operations Research. Proceedings 1977. Edited by R. Henn, B. Korte, and W. Oettli. VI, 270 pages. 1978.

Vol. 158: L. J. Cherene, Set Valued Dynamical Systems and Economic Flow. VIII, 83 pages. 1978.

Vol. 159: Some Aspects of the Foundations of General Equilibrium Theory: The Posthumous Papers of Peter J. Kalman. Edited by J. Green. VI, 167 pages. 1978.

Vol. 160: Integer Programming and Related Areas. A Classified Bibliography. Edited by D. Hausmann. XIV, 314 pages. 1978.

Vol. 161: M. J. Beckmann, Rank in Organizations. VIII, 164 pages. 1978.

Vol. 162: Recent Developments in Variable Structure Systems, Economics and Biology. Proceedings 1977. Edited by R. R. Mohler and A. Ruberti. VI, 326 pages. 1978.

Vol. 163: G. Fandel, Optimale Entscheidungen in Organisationen. VI, 143 Seiten. 1979.

Vol. 164: C. L. Hwang and A. S. M. Masud, Multiple Objective Decision Making – Methods and Applications. A State-of-the-Art Survey. XII, 351 pages. 1979.

Vol. 165: A. Maravall, Identification in Dynamic Shock-Error Models. VIII, 158 pages. 1979.

Vol. 166: R. Cuninghame-Green, Minimax Algebra. XI, 258 pages. 1979.

Vol. 167: M. Faber, Introduction to Modern Austrian Capital Theory. X, 196 pages. 1979.

Vol. 168: Convex Analysis and Mathematical Economics. Proceedings 1978. Edited by J. Kriens. V, 136 pages. 1979.

Vol. 169: A. Rapoport et al., Coalition Formation by Sophisticated Players. VII, 170 pages. 1979.

Vol. 170: A. E. Roth, Axiomatic Models of Bargaining. V, 121 pages. 1979.

Vol. 171: G. F. Newell, Approximate Behavior of Tandem Queues. XI, 410 pages. 1979.

Vol. 172: K. Neumann and U. Steinhardt, GERT Networks and the Time-Oriented Evaluation of Projects. 268 pages. 1979.

Vol. 173: S. Erlander, Optimal Spatial Interaction and the Gravity Model. VII, 107 pages. 1980.

Vol. 174: Extremal Methods and Systems Analysis. Edited by A. V. Fiacco and K. O. Kortanek. XI, 545 pages. 1980.

Vol. 175: S. K. Srinivasan and R. Subramanian, Probabilistic Analysis of Redundant Systems. VII, 356 pages. 1980.

Vol. 176: R. Färe, Laws of Diminishing Returns. VIII, 97 pages. 1980.

Vol. 177: Multiple Criteria Decision Making-Theory and Application. Proceedings, 1979. Edited by G. Fandel and T. Gal. XVI, 570 pages. 1980.

Vol. 178: M. N. Bhattacharyya, Comparison of Box-Jenkins and Bonn Monetary Model Prediction Performance. VII, 146 pages. 1980.

Vol. 179: Recent Results in Stochastic Programming. Proceedings, 1979. Edited by P. Kall and A. Prékopa. IX, 237 pages. 1980.

Vol. 180: J. F. Brotchie, J. W. Dickey and R. Sharpe, TOPAZ – General Planning Technique and its Applications at the Regional, Urban, and Facility Planning Levels. VII, 356 pages. 1980.

Vol. 181: H. D. Sherali and C. M. Shetty, Optimization with Disjunctive Constraints. VIII, 156 pages. 1980.

Vol. 182: J. Wolters, Stochastic Dynamic Properties of Linear Econometric Models. VIII, 154 pages. 1980.

Vol. 183: K. Schittkowski, Nonlinear Programming Codes. VIII, 242 pages. 1980.

Vol. 184: R. E. Burkard and U. Derigs, Assignment and Matching Problems: Solution Methods with FORTRAN-Programs. VIII, 148 pages. 1980.

Vol. 185: C. C. von Weizsäcker, Barriers to Entry. VI, 220 pages. 1980.

Vol. 186: Ch.-L. Hwang and K. Yoon, Multiple Attribute Decision Making – Methods and Applications. A State-of-the-Art-Survey. XI, 259 pages. 1981.

Vol. 187: W. Hock, K. Schittkowski, Test Examples for Nonlinear Programming Codes. V. 178 pages. 1981.

Vol. 188: D. Bös, Economic Theory of Public Enterprise. VII, 142 pages. 1981.

Vol. 189: A. P. Lüthi, Messung wirtschaftlicher Ungleichheit. IX, 287 pages. 1981.

Vol. 190: J. N. Morse, Organizations: Multiple Agents with Multiple Criteria. Proceedings, 1980. VI, 509 pages. 1981.

Vol. 191: H. R. Sneessens, Theory and Estimation of Macroeconomic Rationing Models. VII, 138 pages. 1981.

Vol. 192: H. J. Bierens: Robust Methods and Asymptotic Theory in Nonlinear Econometrics. IX, 198 pages. 1981.

Vol. 193: J. K. Sengupta, Optimal Decisions under Uncertainty. VII, 156 pages. 1981.

Vol. 194: R. W. Shephard, Cost and Production Functions. XI, 104 pages. 1981.

Vol. 195: H. W. Ursprung, Die elementare Katastrophentheorie. Eine Darstellung aus der Sicht der Ökonomie. VII, 332 pages. 1982.

Vol. 196: M. Nermuth, Information Structures in Economics. VIII, 236 pages. 1982.

Vol. 197: Integer Programming and Related Areas. A Classified Bibliography. 1978 – 1981. Edited by R. von Randow. XIV, 338 pages. 1982.

Vol. 198: P. Zweifel, Ein ökonomisches Modell des Arztverhaltens. XIX, 392 Seiten. 1982.

Vol. 199: Evaluating Mathematical Programming Techniques. Proceedings, 1981. Edited by J.M. Mulvey. XI, 379 pages. 1982.

Vol. 200: The Resource Sector in an Open Economy. Edited by H. Siebert. IX, 161 pages. 1984.

Vol. 201: P. M. C. de Boer, Price Effects in Input-Output-Relations: A Theoretical and Empirical Study for the Netherlands 1949–1967. X, 140 pages. 1982.

Vol. 202: U. Witt, J. Perske, SMS – A Program Package for Simulation and Gaming of Stochastic Market Processes and Learning Behavior. VII, 266 pages. 1982.

Vol. 203: Compilation of Input-Output Tables. Proceedings, 1981. Edited by J. V. Skolka. VII, 307 pages. 1982.

Vol. 204: K.C. Mosler, Entscheidungsregeln bei Risiko: Multivariate stochastische Dominanz. VII, 172 Seiten. 1982.

Vol. 205: R. Ramanathan, Introduction to the Theory of Economic Growth. IX, 347 pages. 1982.

Vol. 206: M.H. Karwan, V. Lotfi, J. Telgen, and S. Zionts, Redundancy in Mathematical Programming. VII, 286 pages. 1983.

Vol. 207: Y. Fujimori, Modern Analysis of Value Theory. X, 165 pages. 1982.

Vol. 208: Econometric Decision Models. Proceedings, 1981. Edited by J. Gruber. VI, 364 pages. 1983.

Vol. 209: Essays and Surveys on Multiple Criteria Decision Making. Proceedings, 1982. Edited by P. Hansen. VII, 441 pages. 1983.

Vol. 210: Technology, Organization and Economic Structure. Edited by R. Sato and M.J. Beckmann. VIII, 195 pages. 1983.

Vol. 211: P. van den Heuvel, The Stability of a Macroeconomic System with Quantity Constraints. VII, 169 pages. 1983.

Vol. 212: R. Sato and T. Nôno, Invariance Principles and the Structure of Technology. V, 94 pages. 1983.

Vol. 213: Aspiration Levels in Bargaining and Economic Decision Making. Proceedings, 1982. Edited by R. Tietz. VIII, 406 pages. 1983.

Vol. 214: M. Faber, H. Niemes und G. Stephan, Entropie, Umweltschutz und Rohstoffverbrauch. IX, 181 Seiten. 1983.

Vol. 215: Semi-Infinite Programming and Applications. Proceedings, 1981. Edited by A.V. Fiacco and K.O. Kortanek. XI, 322 pages. 1983.

Vol. 216: H.H. Müller, Fiscal Policies in a General Equilibrium Model with Persistent Unemployment. VI, 92 pages. 1983.

Vol. 217: Ch. Grootaert, The Relation Between Final Demand and Income Distribution. XIV, 105 pages. 1983.

Vol. 218: P. van Loon, A Dynamic Theory of the Firm: Production, Finance and Investment. VII, 191 pages. 1983.

Vol. 219: E. van Damme, Refinements of the Nash Equilibrium Concept. VI, 151 pages. 1983.

Vol. 220: M. Aoki, Notes on Economic Time Series Analysis: System Theoretic Perspectives. IX, 249 pages. 1983.

Vol. 221: S. Nakamura, An Inter-Industry Translog Model of Prices and Technical Change for the West German Economy. XIV, 290 pages. 1984.

Vol. 222: P. Meier, Energy Systems Analysis for Developing Countries. VI, 344 pages. 1984.

Vol. 223: W. Trockel, Market Demand. VIII, 205 pages. 1984.

Vol. 224: M. Kiy, Ein disaggregiertes Prognosesystem für die Bundesrepublik Deutschland. XVIII, 276 Seiten. 1984.

Vol. 225: T.R. von Ungern-Sternberg, Zur Analyse von Märkten mit unvollständiger Nachfragerinformation. IX, 125 Seiten. 1984

Vol. 226: Selected Topics in Operations Research and Mathematical Economics. Proceedings, 1983. Edited by G. Hammer and D. Pallaschke. IX, 478 pages. 1984.

Vol. 227: Risk and Capital. Proceedings, 1983. Edited by G. Bamberg and K. Spremann. VII, 306 pages. 1984.

Vol. 228: Nonlinear Models of Fluctuating Growth. Proceedings, 1983. Edited by R.M. Goodwin, M. Krüger and A. Vercelli. XVII, 277 pages. 1984.

Vol. 229: Interactive Decision Analysis. Proceedings, 1983. Edited by M. Grauer and A.P. Wierzbicki. VIII, 269 pages. 1984.

Vol. 230: Macro-Economic Planning with Conflicting Goals. Proceedings, 1982. Edited by M. Despontin, P. Nijkamp and J. Spronk. VI, 297 pages. 1984.

Vol. 231: G.F. Newell, The M/M/∞ Service System with Ranked Servers in Heavy Traffic. XI, 126 pages. 1984.

Vol. 232: L. Bauwens, Bayesian Full Information Analysis of Simultaneous Equation Models Using Integration by Monte Carlo. VI, 114 pages. 1984.

Vol. 233: G. Wagenhals, The World Copper Market. XI, 190 pages. 1984.

Vol. 234: B.C. Eaves, A Course in Triangulations for Solving Equations with Deformations. III, 302 pages. 1984.

Vol. 235: Stochastic Models in Reliability Theory. Proceedings, 1984. Edited by S. Osaki and Y. Hatoyama. VII, 212 pages. 1984.

Vol. 236: G. Gandolfo, P.C. Padoan, A Disequilibrium Model of Real and Financial Accumulation in an Open Economy. VI, 172 pages. 1984.

Vol. 237: Misspecification Analysis. Proceedings, 1983. Edited by T.K. Dijkstra. V, 129 pages. 1984.

Vol. 238: W. Domschke, A. Drexl, Location and Layout Planning. IV, 134 pages. 1985.

Vol. 239: Microeconomic Models of Housing Markets. Edited K. Stahl. VII, 197 pages. 1985.

Vol. 240: Contributions to Operations Research. Proceedings, 19 Edited by K. Neumann and D. Pallaschke. V, 190 pages. 1985.

Vol. 241: U. Wittmann, Das Konzept rationaler Preiserwartung XI, 310 Seiten. 1985.

Vol. 242: Decision Making with Multiple Objectives. Proceedir 1984. Edited by Y.Y. Haimes and V. Chankong. XI, 571 pages. 19

Vol. 243: Integer Programming and Related Areas. A Classi Bibliography 1981–1984. Edited by R. von Randow. XX, 386 pa 1985.

Vol. 244: Advances in Equilibrium Theory. Proceedings, 1984. Ed by C.D. Aliprantis, O. Burkinshaw and N.J. Rothman. II, 235 pa 1985.

Vol. 245: J.E.M. Wilhelm, Arbitrage Theory. VII, 114 pages. 19

Vol. 246: P.W. Otter, Dynamic Feature Space Modelling, Filte and Self-Tuning Control of Stochastic Systems. XIV, 177 pages. 19

Vol. 247: Optimization and Discrete Choice in Urban Syste Proceedings, 1983. Edited by B.G. Hutchinson, P. Nijkamp M. Batty. VI, 371 pages. 1985.

Vol. 248: Plural Rationality and Interactive Decision Processes. ceedings, 1984. Edited by M. Grauer, M. Thompson and Wierzbicki. VI, 354 pages. 1985.

Vol. 249: Spatial Price Equilibrium: Advances in Theory, Computa and Application. Proceedings, 1984. Edited by P. T. Harker. VII, pages. 1985.

Vol. 250: M. Roubens, Ph. Vincke, Preference Modelling. 94 pages. 1985.

Vol. 251: Input-Output Modeling. Proceedings, 1984. Edited A. Smyshlyaev. VI, 261 pages. 1985.

Vol. 252: A. Birolini, On the Use of Stochastic Processes in Mode Reliability Problems. VI, 105 pages. 1985.

Vol. 253: C. Withagen, Economic Theory and International Trade Natural Exhaustible Resources. VI, 172 pages. 1985.

Vol. 254: S. Müller, Arbitrage Pricing of Contingent Claims. VIII, pages. 1985.

Vol. 255: Nondifferentiable Optimization: Motivations and Appli tions. Proceedings, 1984. Edited by V.F. Demyanov and D. Pallas ke. VI, 350 pages. 1985.

Vol. 256: Convexity and Duality in Optimization. Proceedings, 19 Edited by J. Ponstein. V, 142 pages. 1985.